Praise for

Essential Building Science

Where the heck was Jacob Deva Racusin 40 years ago when I started my career? You could have saved me, and thousands of others, a whole lot of expense and trouble if you had troubled yourself to be born a few decades earlier, and gotten this wonderful gem out to us all a bit sooner. That grouse aside, you have done a masterful — and delightful — job of presenting the complexities of building physics in a way that anyone can understand anduse to their benefit. If you want to build well, folks, read this!

— Bruce King, Structural Engineer and author of *New Carbon Architecture*

This is the book that builders, designer and homeowners have been waiting for... even if they don't know it! In the coming years, understanding building science principles will be consideredas important as drafting skills and driving nails. Jacob Racusin explains everything you needto know in a concise, direct and meaningful way. Anybody who reads this book willimmediately start to make better buildings.

— Chris Magwood, Co-founder and co-director, The Endeavour Centre, and author, *Making Better Buildings*

Jacob Deva Racusin has provided as clear and concise a summary of building science forresidential building as I can imagine. This volume is an invaluable introduction, teaching tool and resource for homeowners, educators and builders. I hope there will be a dog-eared hard copy or an oft-accessed e-version of *Essential Building Science* on every residential building (or renovation) site in North America.

— Tim Krahn, P. Eng.

Getting the building science right in a house — so your home is healthy, durable andenergy efficient — can be a daunting and complex process, but Jacob has distilled this downto principles that are easy to understand. Racusin provides clear guidance for putting these principles into action, including details of building assemblies that work. The contents of this guide are essential for builders and architects, as well as for future homeowners to know what to look for and what to ask for in good design.

— Andrew M Shapiro, Energy Balance, Inc.

Too many of us are suffocating in a stew of mold and toxic chemicals as our buildings disintegrate around us. In this essential book, Jacob Deva Racusin distills the basic tools builders and homeowners need to make efficient homes that start out light on the planet and last a good long time.

— Tristan Roberts, Executive Editor, BuildingGreen.com

essential BUILDING SCIENCE

Understanding **Energy** and **Moisture** in High Performance House Design

Jacob Deva Racusin

New Society Sustainable Building Essentials Series

Series editors
Chris Magwood and Jen Feigin

Title list
Essential Hempcrete Construction, Chris Magwood
Essential Prefab Straw Bale Construction, Chris Magwood
Essential Building Science, Jacob Deva Racusin

See www.newsociety.com/SBES for a complete list of new and forthcoming series titles.

THE SUSTAINABLE BUILDING ESSENTIALS SERIES covers the full range of natural and green building techniques with a focus on sustainable materials and methods and code compliance. Firmly rooted in sound building science and drawing on decades of experience, these large-format, highly illustrated manuals deliver comprehensive, practical guidance from leading experts using a well-organized step-by-step approach. Whether your interest is foundations, walls, insulation, mechanical systems or final finishes, these unique books present the essential information on each topic including:

- Material specifications, testing and building code references
- Plan drawings for all common applications
- Tool lists and complete installation instructions
- Finishing, maintenance and renovation techniques
- Budgeting and labor estimates
- Additional resources

Written by the world's leading sustainable builders, designers and engineers, these succinct, user-friendly handbooks are indispensable tools for any project where accurate and reliable information is key to success. GET THE ESSENTIALS!

Cover design by Diane McIntosh.
Interior illustrations © Dale Brownson.
Thumbs up art: AdobeStock_23490949.
Chapter banner: Ace McArleton — New Frameworks Natural Design/Build.
Wood background: AdobeStock_65301138

Printed in Canada. Second printing, June 2021.

Any other inquiries can be directed by mail to:
New Society Publishers
P.O. Box 189, Gabriola Island, BC V0R 1X0, Canada
(250) 247-9737

Library and Archives Canada Cataloguing in Publication

Racusin, Jacob Deva, author
Essential building science : understanding energy and moisture in high performance house design /Jacob Deva Racusin.

(Sustainable building essentials)
Includes bibliographical references and index.
Issued in print and electronic formats.
ISBN 978-0-86571-834-0 (paperback).--ISBN 978-1-55092-629-3 (ebook)

1. Dwellings--Thermal properties. 2. Dwellings--Heating and ventilation. 3. Dwellings--Energy conservation. 4. Dwellings--Design and construction. 5. Dampness in buildings. I. Title. II. Series: Sustainable building essentials

TH9031.R32 2016 693.8'9 C2016-906214-7
C2016-906215-5

Funded by the Government of Canada | Financé par le gouvernement du Canada | Canada

New Society Publishers' mission is to publish books that contribute in fundamental ways to building an ecologically sustainable and just society, and to do so with the least possible impact on the environment, in a manner that models this vision.

Contents

Acknowledgments

THIS BOOK WOULD NOT BE POSSIBLE without the incredible support of a community of passionate and committed individuals and organizations. I offer my deep and humble gratitude to:

Ace McArleton, Ben Graham, and the New Frameworks team, for their fearless pursuit of a better world.

Yestermorrow Design/Build School, for their commitment to empowering our society to design/build great things.

The global natural building community, for believing in a path to deep sustainability.

The global building science community, for lighting our way to a deeper understanding of our world.

New Society Publishers, for bringing this and so many vital publications into the hands of those who need them.

Chris Magwood and Jen Feigin, for their tireless support and clear vision of the potential held in this book.

My technical editors, Robert Riversong, Andy Shapiro, and Bruce Courtot, whose brilliant minds and efforts helped ensure the quality and accuracy of this work.

My many teachers and mentors in this work, on whose shoulders I stand every day.

My family, Mary, Elijah, Mimi, and Micah, and Bob, Sharon, and Jess, for absolutely everything.

This book is dedicated to those who are dedicated to those who come after them.

Introduction

By means of microscopic observation and astronomical projection
the lotus flower can become the foundation for an entire theory of the universe
and an agent whereby we may perceive Truth.

— Yukio Mishima

A Case for Building Science

WHY DO WE NEED BUILDING SCIENCE? We've been building shelter just fine, all around the world and throughout time, without quantifying the physics of heat and moisture movement. Safe shelter is a birthright to all, and the potential to design, to build, to create, exists within each of us. Doesn't all this complicated building science just get between us and the work, complicate things and distract us from our intuition?

So, why do we need building science? Because our buildings have grown increasingly complex, and we expect very high levels

Fig. I.1

of comfort from them. Many of the biggest problems we have with our homes arise from problems with how our buildings manage heat and moisture. Much of the housing stock in North America is plagued by significant moisture problems (rot and mold), poor indoor air quality, and exorbitant energy loads, leading to great expense — for both the climate and for owners. We ask a lot of our buildings in terms of interior climate control, comfort, healthy space, and good durability, and in pursuit of these goals we burn a lot of carbon and use a lot of toxic materials. We need the tools and understanding afforded to us by the practice of applied building science to understand these problems better, so we can choose solutions that will create buildings that meet our needs without enacting such a heavy toll on the planet's well-being.

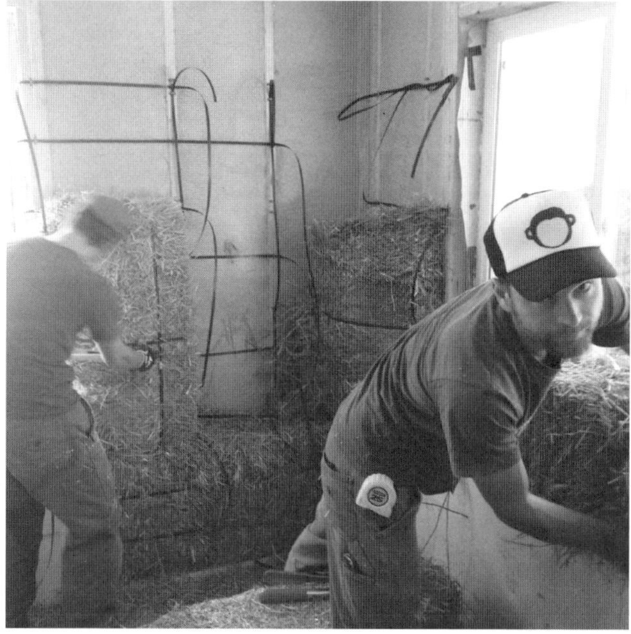

Fig. I.2: *A return to time-tested natural materials is a consequence of many years of building with toxic materials.* CREDIT: ACE MCARLETON — NEW FRAMEWORKS NATURAL DESIGN/BUILD.

Some of these solutions are old. We have been building with low-impact materials such as wood, stone, earth, lime, and straw for thousands of years. After nearly a century of building with whatever (possibly toxic) materials manufacturers were offering, many people now demand healthy homes built with low-toxic materials that provide high levels of efficiency with low carbon footprints. Those time-tested natural materials we built with for all those pre-Industrial Revolution years are more relevant than ever. To adapt them into this new context, we need to apply building science principles to ensure we use these natural materials effectively to create durable, high-performance homes.

Some of these solutions are new. So, we need to be able to predict how a building might perform before taking risks and trying new things. We can never really know, of course, until a building has been up for a few decades — but we need to know enough to manage risk and innovate appropriately. We need to have some benchmarks and goals to know if the new things we try are indeed working.

Most of all, we need building science because we need to elevate the bar of quality for our built environment, not just incrementally, but significantly and quickly, to address the critical issues of climate destabilization, unsafe and insecure housing, "sick" buildings, and housing that costs too much to operate. We can use what we have learned through observation and experimentation with heat, moisture, and materials, and by referencing patterns that have surfaced in thousands of buildings over the last few hundred years to dramatically improve the durability, safety, comfort, cost, and impact of our buildings. While it is far from the only discipline involved in creating the built environment, building science has to play a critical role if we want to achieve the lofty goals we have set for ourselves.

Building Science: A Very Brief History

Humans are innate scientists — or at least, there are a few in every crowd. A true survey of the applied practice of building science, across cultures and throughout the ages, would make for a fascinating read. Here in North America, we have been building wood-framed and masonry (earth, brick, stone) homes for the last few hundred years. Most have been lost to the ravages of time, but many have endured. Homes built of solid wood (i.e., logs, planks) and earth were very durable and quite comfortable. As structural engineering advanced, and we began building lighter structures out of smaller things (wood studs, steel, glass), our houses became less comfortable as we lost the thermal properties and tightness of those solid earth and wood walls. Around the 1920s and 1930s, we decided as an industry that insulation in our buildings was a good idea and worthy of bringing to market.

It took only a short decade or two before we began to see the first signs of moisture damage — paint began to peel from the siding. The prevailing hypothesis: vapor was moving through the walls from the inside, pushing through the wood siding, and driving the oil-based paint off the wood. The proposal, codified in building codes for decades to come: a vapor barrier to be installed on the inside of the building, stopping the vapor before it could move into the wall. The result: the continuation of vapor-related damage on both exterior and interior wall surfaces. The diagnosis was correct, but as we will learn, simply putting up a vapor barrier does not take into account the full cycle of vapor movement in our

Fig. I.3: *Building physics is at play whether we understand it or not — the penalty for ignoring the physics can be severe!* Credit: Top and bottom right: New Frameworks Natural Design/Build.

buildings; therefore, it is not in and of itself a suitable cure to the problem. A systems-level solution is needed for a systems-level problem.

In the 1970s, the United States and Canada experienced dramatic increases in fuel prices, resulting in energy improvement measures across all sectors of the economy, including buildings. We developed diagnostic technology, like the blower door, to be able to understand how air leaks occur in our buildings, and we dramatically increased the amount of insulation in our walls and roofs. New materials were employed; new innovations were brought to the market.

Again, we began to see moisture damage: rotting roofs, sheathing, siding, and framing. Over the decades to follow, we have steadily been learning what we can and cannot get away with in our buildings — as the invisible worlds of heat and vapor are made visible through the observations of how our buildings hold up, and what patterns of failure emerge. When we add more insulation, walls get colder and stay wet longer. Some materials seem to hold up when they get wet, and others fail quickly. For those of you new to the high-performance building world, it is worth noting that few of the strategies we will present are new — we have been building super-insulated, airtight, low-load, passive solar-heated buildings successfully (and not so successfully) for over 40 years. It is on the shoulders of the giants who raised these buildings that we stand, and the presentation of the information in this book is possible only because of the curiosity, leadership, fearlessness, and innovation of the designers, builders, engineers, and owners before us who were committed to learning what they could in pursuit of building better buildings.

Fig. I.4: *Proof of concept for high performance/ passive solar construction was established in 1939 with groundbreaking homes such as Keck and Keck's first known "solar home" in Glenview, IL.* CREDIT: NICOLE SERRADIMIGNI — VHT STUDIOS.

How to Use This Book

When I started designing and building my house in 2000, having neither practical nor academic experience, I was compelled to learn as much as possible as quickly as possible. Further complicating matters was my choice to build with timbers, straw bales, and earthen plaster, about which little was known and much was misunderstood, particularly in the context of the cold, wet climate of the northern Vermont mountains. Having seen how quickly moisture can break down a conventional building, and knowing what a bad reputation building with earth and straw had in my region, I wanted to know all I could about how these materials worked with moisture and, as an assembly, how these walls performed in more extreme climates. Fortunately, I was able to draw upon work from researchers such as Dr. John Straube and Don Fugler, engineers such as Marc Rosenbaum and Bruce King, and builders such as Paul Lacinski and Chris Magwood (and many others), who had already started to develop some regional best practices based on fundamental building science principles.

This book is intended to be the resource I wish I had when I started this journey, and it is written with the motivated or curious homeowner, aspirational or active owner-builder, and developing or practicing builder or architect in mind. This book serves as a foundational approach to applied building science — the essential rationale, physics principles, and strategies for building a high-performance home. The focus is restricted to energy and moisture, although related topics such as structure, acoustics, and lighting are referenced throughout the book.

This book won't tell you everything you need to know about how to build a high-performance

Fig. I.5: *Thanks to our ever-growing understanding of how buildings work and respond to their environments, today we are designing and building some of the best-performing homes that have ever existed.*

Credit: Top and bottom: Ace McArleton — New Frameworks Natural Design/Build.

house. It will, however, lay out a systems-based approach to developing critical thinking that you can apply to your home. This book will arm you with the information and strategies necessary to think your way through decisions regarding materials, assemblies, and mechanical systems, and it will help identify the areas of further study or greater expertise that may be required to make good decisions. Finally, this book will create a framework for asking better questions, which in time and with practice lead to better answers, and ultimately to better buildings.

Part I:

Rationale

Credit: New Frameworks Natural Design/Build.

Chapter 1

Establishing Goals

BEFORE WE DIVE INTO THE THEORY, strategies, and details of how to manage heat and moisture in our buildings, it is critical that we begin by establishing the values and goals for our building projects, knowing that every project, with different stakeholders and serving specific purposes, will have a unique composition of values and goals. Why is this so important? After all, the physics are the same; heat and moisture don't change based on your proclivities. The importance lies in being able to ensure that the many decisions you will have to make in response to the guidelines presented in this book all lead toward a result that matches your original intentions. This is not something unique to building science — this is an important approach for any project. It is all too easy to think that there is a "right way" from a building science perspective, that just needs to be identified. Rather, the information concerning the performance of our buildings can be used in support of our goals — or it can lead us away from them if we haven't done a good job of setting clear objectives.

There are many definitions and graphical flowcharts representing this process. Here is a simple structure that runs through the exercise with a few basic examples:

Identify Values

First, what values motivate this project and what are your priorities? It is very important to note that you will have values that will conflict with each other; you will have to assign a hierarchy of importance to them. The more time you can spend getting clear about your priorities (and practicing how to communicate them clearly to others), the easier it will be to make decisions to resolve inherent conflicts — many of which inevitably involve cost — when they surface.

Define Goals

Next, what are the specific goals and results you are working toward? The goals should reflect your values. Set these goals early in the design process to guide the strategies and details.

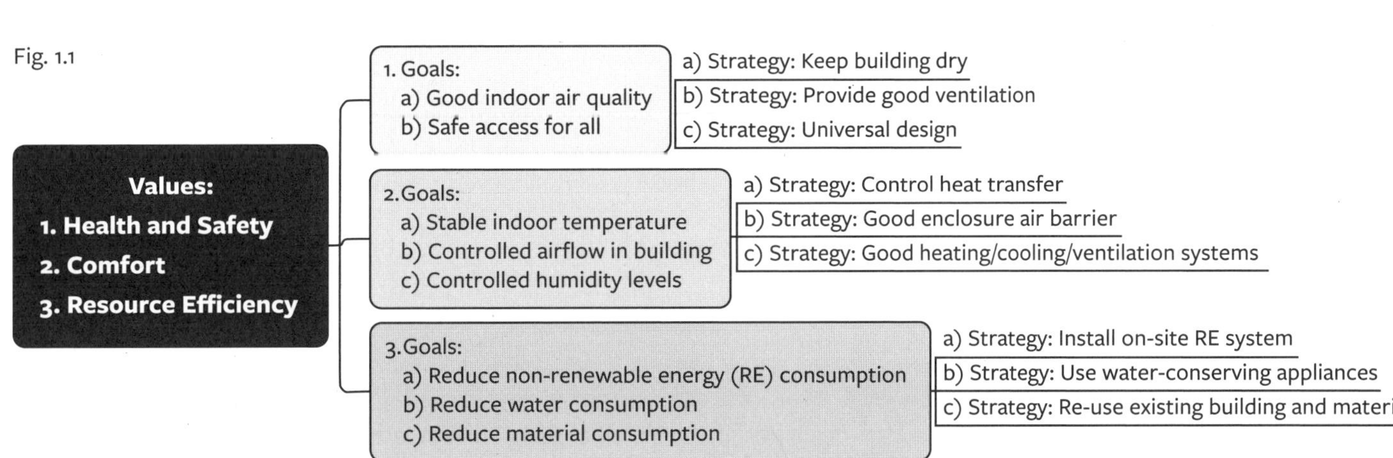

Fig. 1.1

Table 1.1

Value: Health and Safety	
Goal: Good indoor air quality	
Strategies	***Actions***
Keep building dry (avoid mold)	Follow moisture control strategies to create a well-detailed enclosure that avoids leaks, condensation, and groundwater intrusion.
Good ventilation	Install a well-sized and appropriately designed ventilation system for the home, and ensure it runs correctly.
Reduce pollutants	Avoid products with off-gassing chemicals such as the solvents and glues common in paints, floor covering, and furnishings.
Combustion safety	Avoid the use of combustion appliances and/or use sealed combustion appliances when necessary.
Goal: Safe Use for All	
Universal Design (UD)	Follow established UD guidelines to ensure access to those needing assistance moving about in the home.

Table 1.2

Value: Comfort	
Goal: Stable indoor temperature	
Strategies	***Actions***
Control heat transfer	Follow thermal control strategies for a well-detailed, airtight and well-insulated enclosure, and include thermal mass inside the building.
Provide appropriate heating and cooling	Install a well-sized and appropriately designed heating/AC system for the home, and ensure it runs correctly and is serviceable.
Keep surface temperatures warm	Insulate slabs, walls, and ceilings. Install thermally efficient windows.
Goal: Controlled air flow (no drafts when cold, circulation when warm)	
Install quality air barrier	Ensure air barrier detailing in design, and perform quality control in the field with a blower-door test.
Good ventilation	Install a well-sized and appropriately designed ventilation system for the home, and ensure it runs correctly.
Goal: Controlled humidity levels	
Same as for other goals addressing comfort	Keep building dry, keep surfaces warm, install an appropriate ventilation system.

Setting measurable targets will make it easier to identify strategies and understand the results. Examples might include wanting a net zero energy home, or avoiding the use of any of the materials on the "Living Building Challenge Red List."

Develop Strategies

How can you reach your goals? You will need to decide on the steps and solutions that are likely to give you the results you seek. The strategies laid out in this book that support good moisture and thermal control may be valid unto themselves, but depending on the other goals you have set for this building, some of these strategies will be highly relevant, and others less so.

This is a deeply personal exercise, whether for owners or for professionals. The more intention you can bring to this process, the more likely it is that the final project will reflect your values, reach your goals, and employ successful strategies. In the world of high-performance building, common priorities emerge; some of these appear in the list below to help you seed the process of establishing your own building priorities.

Health and Safety

While this may seem like an obvious standard for all buildings, it is quite clear from the preponderance of newly constructed homes with terrible indoor air quality (IAQ) that this is not a universally held priority.

Comfort

Comfort is driven by humidity and temperature. Temperature here is not just air temperature, but *operative temperature,* which is a weighted average of air temperature, surface temperatures of a space, and air velocity. In high-performance homes, good insulation keeps surface temperatures closer to air temperatures, enhancing comfort.[1]

Durability

This is a priority often taken for granted, but rarely understood, thus keeping the renovation industry employed for the foreseeable future. We explore strategies for moisture durability in depth in this book — this is perhaps the governing focus of building science. One of the crucial questions to answer is this: "How to we keep our buildings really comfortable using little energy and without them rotting prematurely?"

Resource Efficiency

Resource inputs into the building — materials, water and energy — must be managed to achieve goals relating to affordability, independence, resiliency, and ecological and social responsibility — this is one priority that can unite people across the political spectrum!

Fig. 1.2: *Sweating windows can be caused by poor quality windows, elevated indoor humidity levels, or a combination of the two. They are a sign that the building's durability may be compromised in other parts of the enclosure as well.* Credit: New Frameworks Natural Design/Build.

Table 1.3

Value: Durability	
Goal: Keep rain and snow from getting in	
Strategies	***Actions***
Control bulk water entry	Follow moisture-control strategies of a well-detailed enclosure, avoiding leaks and groundwater intrusion.
Control surface water	Install storm water control (i.e., gutters) that integrates with a well-drained site.
Goal: Keep condensation out of the enclosure	
Install quality air barrier	Ensure air barrier detailing in design, and perform quality control in the field with a blower-door test.
Appropriate ventilation system	Install a well-sized and appropriately designed ventilation system for the home, and ensure it runs correctly.

Table 1.4

Value: Resource Efficiency	
Goal: Reduce non-renewable energy consumption	
Strategies	***Actions***
Control heat transfer	Follow thermal control strategies of a well-detailed air-tight and well-insulated enclosure, avoiding thermal bridging.
Provide appropriate heating and cooling	Install a well-sized and appropriately designed heating/AC system for the home, and ensure it runs correctly and is serviceable.
Install renewable energy systems	Install solar electric or thermal panels; use biomass for solid-fuel heating.
Goal: Reduce water consumption	
Use water-conserving appliances	Install low-flow faucets, showerheads, and toilets, and service regularly to eliminate leaks.
Integrate water-conserving systems	Design for rainwater catchment, greywater recycling, and composting toilet systems.
Goal: Reduce material consumption	
Re-use existing buildings and materials	Rehab existing buildings when possible; salvage what must be demolished; purchase quality used materials when possible.
Use low-impact materials	Use natural and regionally sourced abundant materials such as stone, wood, straw/hemp; select products with high recycled content (everything from steel beams to cellulose insulation).

Embodied Carbon and Operational Carbon

There is a particular urgency regarding the climate impact of our buildings. At the time of this writing, 97% of the science community has come to consensus about the fact that anthropogenic (human-caused) greenhouse gas emissions will lead to irreversible climate disruption if dramatic reductions are not realized within the first few decades of this century.[2] In order to address this emergency, Architecture 2030, a leader in the climate action movement in the building industry, has called for zero carbon emissions by 2050 for all new construction. This includes both *operational carbon,* which is the carbon load created by the use of energy to heat and power a building, as well as *embodied carbon,* which is the carbon that is released in the manufacturing, production, and transportation of our building materials.[3]

For the decades that the building community has been working on improving energy efficiency, the strategy has been to use relatively high-embodied carbon materials, such as insulation, to offset long-term operational carbon loads, which over the life of the building are considerably greater (especially for conventional buildings). However, this is changing rapidly, and the building community must now take the consideration of embodied carbon very seriously, for the following reasons:

- Within the short time frame we have to reduce impacts, using high-embodied-carbon insulation to reach high levels of building energy performance, especially spray and board foam, may not save enough energy to justify its use (as compared to building a lower-performance building). No longer can we "front-load" 20 or 30 years' worth of operational carbon into the construction of our buildings and wait for the long-term reductions to kick in — we need immediate reductions, which must come from the construction phase, particularly in material selection.
- As we build buildings with lower and lower operational loads, a higher percentage of our carbon impact is in our materials. We know how to improve energy efficiency, but we must focus on how to do it using lower-carbon ⟶

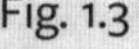
Fig. 1.3

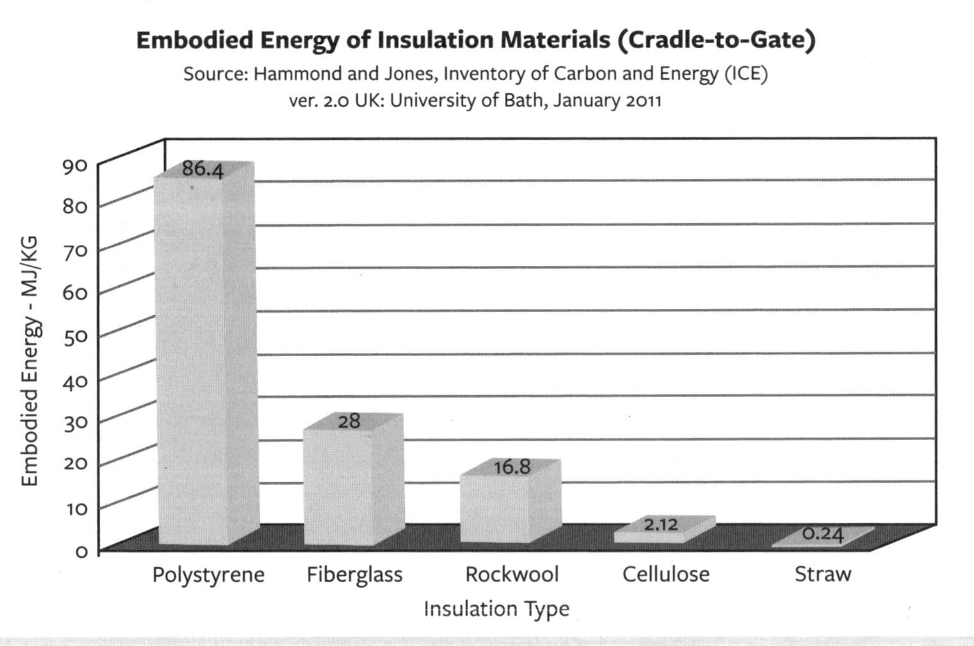

materials and practices. "Net Zero" and other approaches using renewable energy to offset operational carbon does nothing to offset our embodied carbon!

- We now have the technology to measure, evaluate, and effectively use low-carbon materials to build high-performance houses. Many of these materials are some of the oldest materials we've built with — wood, clay, stone, straw. We now have the experience, scientific understanding, testing and evaluation technology, and modeling software and datasets to put these materials into a modern context.
- Rather than a future vision, the low-carbon high-performance home is both a practical reality and an urgent need. When setting your values, goals, and strategies, we encourage you to hold climate action as a value, zero-carbon building as a goal, and using low-carbon materials as a strategy!

Socio-ecological Equity

This term may not be familiar to everyone. In recognizing that what is good for the planet is generally also good for people, and that negative environmental impacts directly relate to negative social impacts often borne by the most vulnerable of members of our population, we must consider both the social and ecological impacts of our buildings as one and the same. This issue is important to address, for several reasons: 1) these are often "invisible" issues, as the damage may be inflicted many miles away from the building site, either prior to or after the building's useful service life; 2) there is not always a built-in financial or other direct incentive experienced by the stakeholders prompting them to make decisions in support of socio-ecological responsibility, so a strong intention to keep these priorities visible is required; and, 3) in many cases — especially regarding global climate change — there is the need for immediate action to avoid catastrophic results such as species extinction, flooding and droughts, or irreversible climate destabilization. Every building can potentially either exacerbate these problems or provide solutions. If we truly take seriously the perspective of building science in the development of our buildings, we must consider a broader context of applied science that not only evaluates how heat and moisture work upon our buildings, but how the development of our buildings work upon our social and ecological support systems.

If we truly take seriously the perspective of building science in the development of our buildings, we must consider a broader context of applied science that not only evaluates how heat and moisture work upon our buildings, but how the development of our buildings work upon our social and ecological support systems.

Table 1.5

Value: Socio-ecological Equity	
Goal: Reduce atmospheric carbon loading	
Strategies	***Actions***
Same as for Resource Efficiency	Same as for Resource Efficiency
Reduce embodied carbon of building	Choose low-carbon building materials and practices.
Goal: Reduce fossil-fuel consumption	
Same as for Resource Efficiency	Same as for Resource Efficiency
Goal: Reduce toxic footprint	
Same as for Resource Efficiency	Same as for material Resource Efficiency.
Use non-embodied-toxic materials	Select materials that have a reduced industrial toxic footprint, avoiding petrochemicals, intensively mined and processed materials, and chemical solvents.

Resiliency

As we continue to move through the 21st century, we find an increasing value placed on buildings that can respond to both incidental disruptions caused by specific events (i.e., storms, droughts), as well as long-term disruptions caused by a changing energy landscape, climate patterns, and resulting social and political impacts. To that end, there has been much discussion in the green building world about *resilient design,* focusing on buildings that can adapt to the known and unknown changes and stresses of the century ahead — and thrive.

Fig. 1.4: *For homes run on critical electric systems, a stand-by generator may be relevant, particularly in rural locations or areas where grid power is less reliable.* CREDIT: NEW FRAMEWORKS NATURAL DESIGN/BUILD.

Making a Plan

Clearly, many strategies will support multiple goals and values — a great example is the strategy of building a high-performance enclosure. Accordingly, we devote the majority of the rest of this book in service to this approach. Having identified this, however, the design details of this enclosure, as balanced against cost, time frame, and other practical logistics, will vary greatly based on your own hierarchy of values and the goals you choose to set. So, take some time to get clear on what it is you want, and let us now begin to explore the strategies available to us.

Table 1.6

Value: Resiliency	
Goal: Simple, reliable, adaptable systems	
Strategies	***Actions***
Provide appropriate HVAC systems	Ensure efficient systems are easy to repair, not reliant on specialized labor, tools, or parts.
Design for adaptability	Allow for the ability to modify mechanical, electric, plumbing, and other services, as well as efficient remodels without compromising integrity of the enclosure.
Goal: Building functionality with service loss	
High-performance enclosure to provide stable interior climate	Follow thermal control strategies of a well-detailed air-tight and well-insulated enclosure, to ensure safe living conditions in the event of power or fuel disruption and resulting HVAC system downtime.
Backup systems for critical functions	Employ backup systems and power supplies to allow critical buildings functions (i.e., HVAC, water supply, refrigeration, communications) to continue by availability of multiple energy sources.
Goal: Community support	
Community integration	Integrate associated food, transportation, energy, and social systems with local and regional communities to support both the operation and function of the building, as well as the social and service needs of the owners. Examples include neighborhood tool sharing, community gardens, and community disaster-response action plans.

Part II:

Fundamentals of Building Physics

Credit: Ace McArleton — New Frameworks Natural Design/Build.

Chapter 2

Thermal Dynamics: Understanding Heat Loss and Gain

In this chapter, we'll examine the underlying dynamics of the movement of heat to help guide our approach to efficiently creating comfortable conditions in our buildings.

Understanding how heat is gained and lost in a building begins with a review of Newton's first law of thermodynamics, known as the law of conservation of energy. The simplified version of this law states that energy cannot be created or destroyed — it is simply transformed. It is fundamentally important to remember that energy doesn't "go away" or "disappear" — it can't! Our concern is *heat transfer* or *transformation* — how energy moves from one place or form to another. Starting here will help guide our conversation.

Newton's second law is the law of entropy. The simplified version of this law says, in part, that heat moves from an area of greater concentration to one of lesser concentration; this

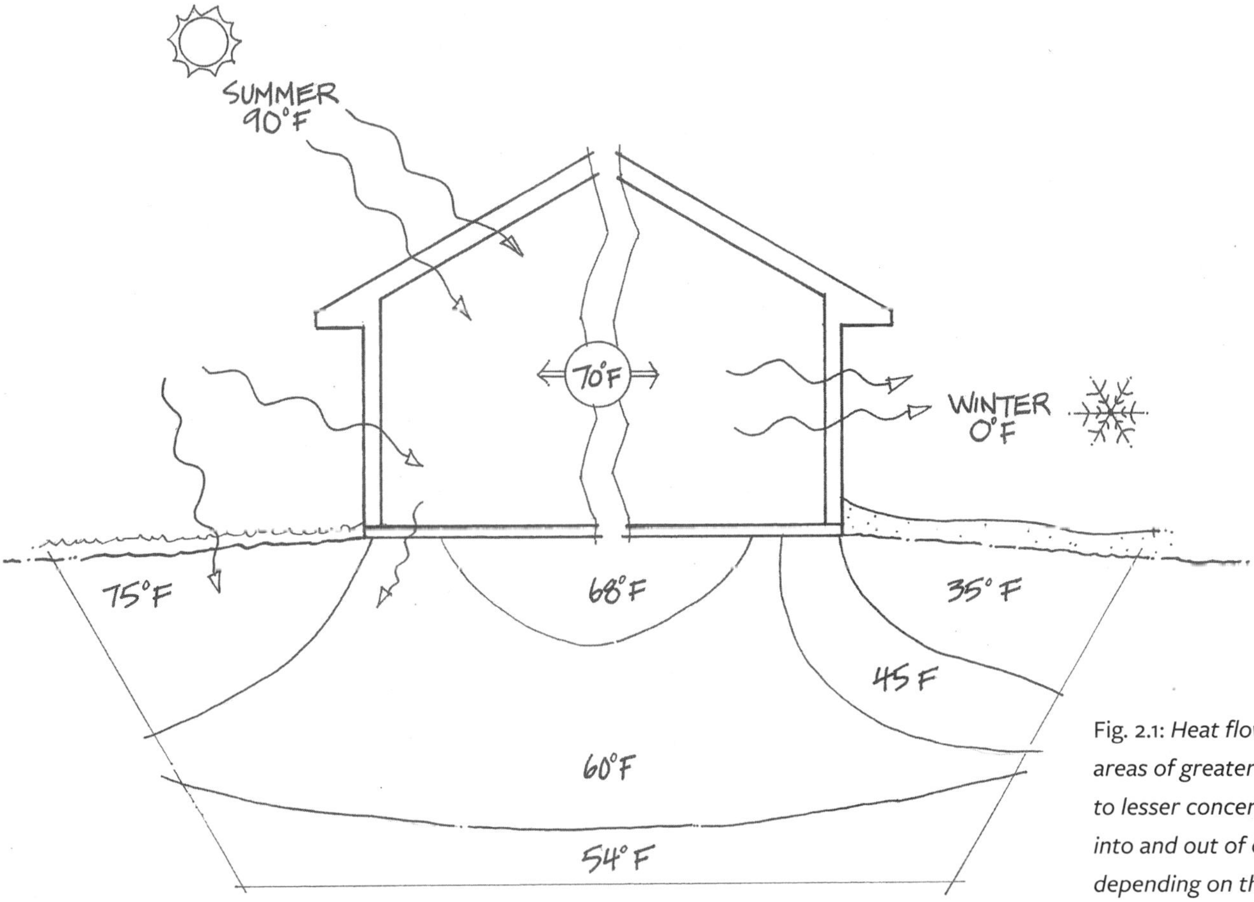

Fig. 2.1: *Heat flows from areas of greater concentration to lesser concentration, both into and out of our homes, depending on the weather.*

is referred to as the *concentration gradient.* Why does the house get cold when the fire goes out or hot when the sun comes out? Energy is moving along a concentration gradient in an act of true democracy, the equal distribution of energy. (Entropy doesn't only relate to heat. I have often thought of my children as terrific forces of entropy: shifting the complex ordering of the belongings in our house to a state of evenly distributed disorder.)

This concentration gradient will have an effect on the direction of heat flow. We can classify the pattern of heat flow in two ways: *steady-state,* where the heat is flowing steadily in one direction (say, from the sun to the earth), and *transient,* when the direction of heat flow changes (between the inside and outside of a building, changing from hour to hour and season to season). When trying to understand how the heat moves in our buildings, we must always be sure to evaluate which direction the heat is flowing. But whether steady or transient, remember that heat is always trying to reach a state of equilibrium, flowing from areas of high concentration to low.

One more fundamental concept to consider is the differences among energy, heat, and temperature:

- *Energy* is the measure of the ability of a physical system to enact force upon another physical system.
- *Heat* is one form of energy, with frequencies in the infrared spectrum.
- *Temperature* is a measurement of the intensity, not the amount, of heat.

Though we often equate *heat* and *temperature,* they are not the same. Different materials can be of the same temperature, yet hold drastically different amounts of heat. It takes 62.4 British Thermal Units (BTUs — a measurement of heat energy) to raise one cubic foot of water one degree Fahrenheit. It would take only 20.41 BTUs (less than a third) to raise a cubic foot of wood the same one degree. A thermometer may read the same temperature for the water and the wood, but the amount of heat energy held in the material is not the same.

Fig. 2.2: *To most effectively slow unwanted heat gain or loss through our buildings, a whole-systems approach is taken to address all forms of heat transfer. Here, a vapor-variable interior air barrier contains a deep cavity of cellulose insulation. The air barrier is outboard of the "service cavity," insulated with cotton batts, where the utilities are run.*
CREDIT: ACE MCARLETON — NEW FRAMEWORKS NATURAL DESIGN/BUILD

Methods of Heat Transfer

Sensible heat is that which can be measured by a thermometer; it moves in three ways: radiation, conduction, and convection. Most often, we'll find that heat is moving in more than one of these ways at once, but one is often more dominant than the others. Understanding these different methods of heat transfer will instruct us on how best to keep the heat from leaving or entering a building unnecessarily, as well as how best to transfer heat to occupants — after all, comfort is a primary objective in heating a building, and operative temperature is informed by both radiant and convective heat, as well as air velocity.

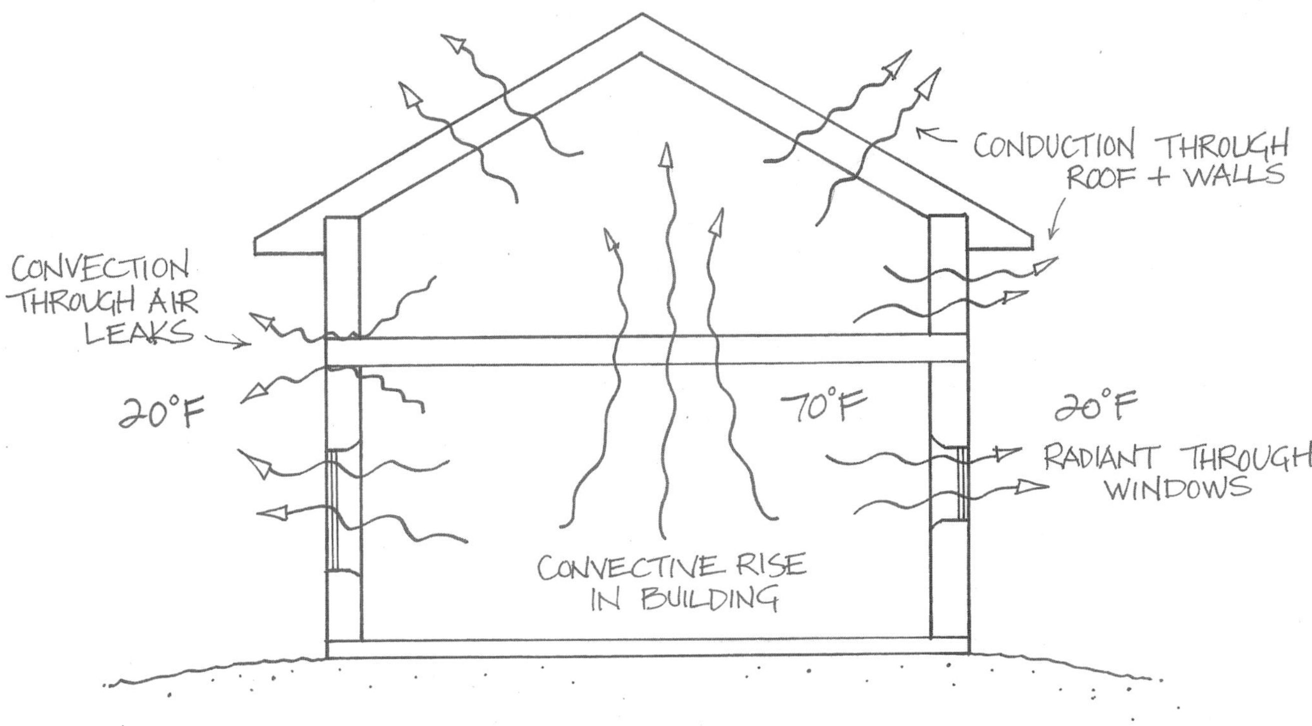

Fig. 2.3: *Heat enters, leaves, and moves within our buildings in many different ways, but only through three different methods: conduction, convection, and radiation.*

Radiation

Radiation is the simplest example of Newton's second law: it is heat moving across a concentration gradient through air or space. Radiant energy moves in all directions equally, but it only moves in straight lines. If a warm object is in your line of sight, it can heat you by radiation. The sun is a great example: heat emitted from the sun travels directly to the earth through space, and to our homes and gardens and bodies through the air.

Radiation is a function of emissivity (the ability of a surface to emit infrared energy) and temperature to the fourth power, magnifying its impact. As an equation, the radiation energy of a material (Q) would be represented as:

$$Q = \sigma \times e \times T^4$$

Where σ is a known constant (the symbol is the lowercase sigma), *e* is the emissivity of the material, and T is temperature in Kelvin (note that temperature is an exponential value in this equation, having a large impact on the radiation potential).

When radiant energy strikes a material, energy is either reflected (bounces off), absorbed (held within), or transmitted (passes through). Energy hitting an opaque material will either bounce off or be absorbed. Kirchhoff's Law tells us that, at thermal equilibrium, the emissivity of a body (or surface) equals its absorptivity. So, if a material can absorb lots of energy — high absorptivity — it can radiate lots of energy — high emissivity. Shiny metal is highly reflective (will bounce heat off of its surface), and has very low emissivity, meaning it will not radiate its heat very effectively; this is why foil-faced insulation acts as a radiant barrier if properly installed. As a heat source, the higher a material's thermal capacity, the more energy it will have to radiate; this is why many radiant heaters use high-capacity materials such as stone or water.

In buildings, we see properties of emissivity and thermal capacity controlled in different areas — shiny surfaces to keep heat from radiating into an attic; massive earthen walls to keep excessive heat from moving into a building; concrete hydronic floors or stone heaters slowly and steadily emitting heat into a room.

It is important to understand that radiant barriers reflect radiation only when there is an air gap on at least one side of the shiny surface; if there is no gap, the heat transfer occurs through conduction (more below) and the benefits of the radiant barrier are lost entirely.[1]

When radiation is not controlled effectively, the results can be extreme surface temperatures and discomfort, unwanted heat loss or gain through window and door glazing, or overheated attics.

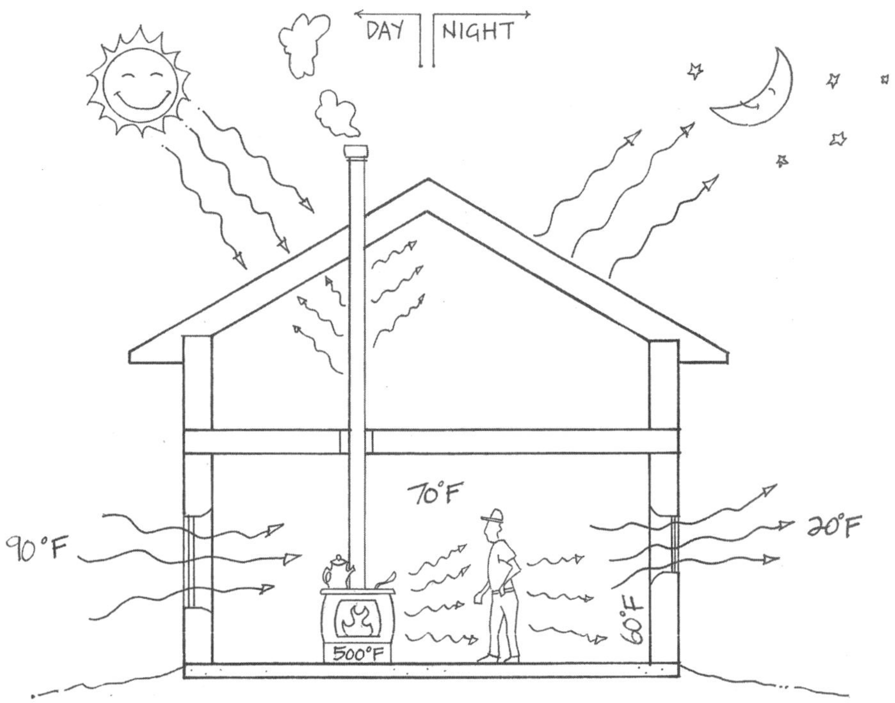

Fig. 2.4: *Radiant heat loss and gain can occur in different ways: through window glazing, from an occupant to a cool surface, and from a heater to an occupant.*

Conduction

Conduction is the transfer of heat energy directly through a material. As heat enters one surface of the material, the increased energy excites the molecules in the material. As these excited molecules, in turn, excite their neighbors, the heat moves through the material. The potential for this form of heat transfer is based upon the material's conductivity, its thickness, and its surface area. Represented in an equation, known as Fourier's Law, it looks like this:

$$Q = K/L \times \Delta T \times A$$

Where Q is the heat rate, K is the coefficient of conductivity of the material, L is the thickness, ΔT is the difference in temperature across the material, and A is the area of the material. Let's examine one variable at a time.

Coefficient of conductivity:

This is simply a measure of how well a material allows heat to travel through it. A high coefficient would signify a good *conductor,* such as metal, whereas a low coefficient would signify a good *insulator,* such as straw. The less dense a material, the less conductive it tends to be; still air is a great insulator due to its lack of density. In identifying a good insulator (of great importance in a building enclosure), we look for a material that entrains a lot of small pockets of still air. Looking at a straw bale, it is the entrained air within and around the hollow straws that supports its performance as an insulator.

Thickness:

As the denominator of the equation, the thicker the material, the lower the conductive potential. This is why most cold-climate building codes require 2×6 studs rather than 2×4 — not for structure, but to increase the thickness of the insulation to reduce conductive losses in the building. This is also why super-insulation strategies rely on deeper wall and ceiling cavities to accommodate increased insulation thicknesses.

Temperature difference (ΔT):

The greater the temperature gradient, the greater the heat transfer. This explains why we are so much more concerned with insulation detailing in cold climates, where our ΔT can be 100°F (55°C) in the depths of winter — that's a lot of drive for conductive loss.

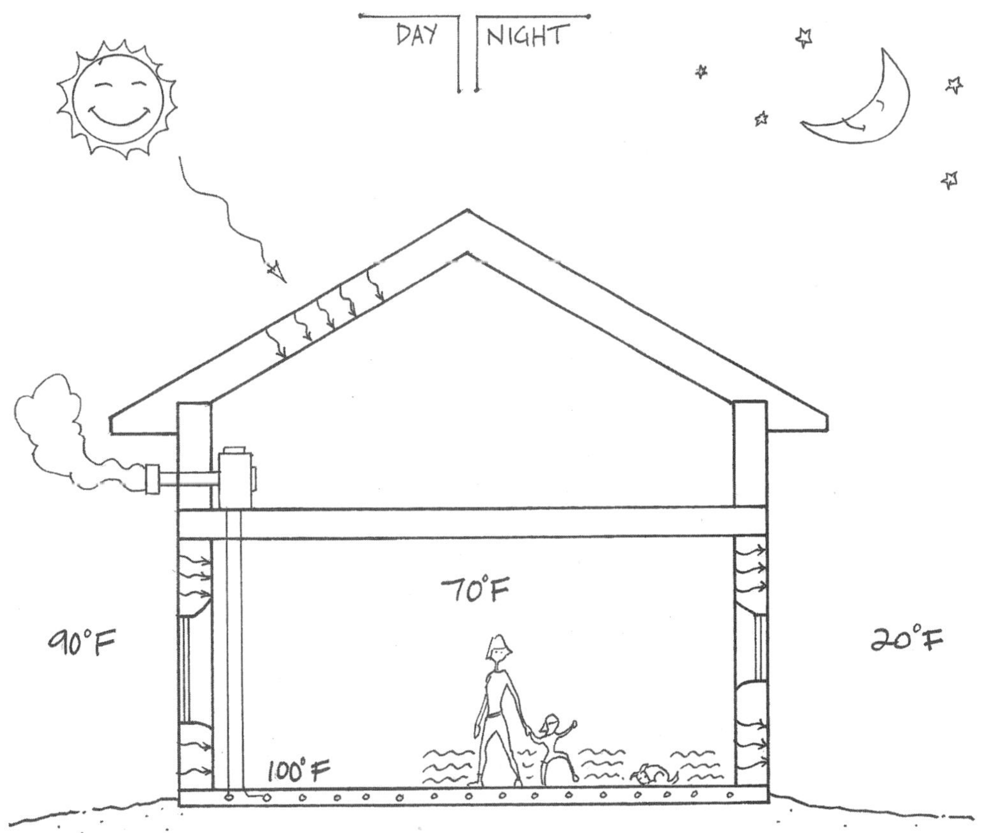

Fig. 2.5: *Unwanted conductive heat loss or gain through framing can significantly drive up energy consumption in a home, while conduction from a hydronic floor into our feet is lovely in the winter.*

Area:

The greater the area of exposure, the greater the potential conductive heat loss — smaller building enclosure, less loss. This also speaks to the role of conductive materials in the wall, such as framing members; we'll discuss this point in greater detail, below.

Measuring Conductivity

The conductivity of building materials is expressed as a U-value, which measures a material's conductive heat transmittance rate. R-value, being the inverse (R = 1/U), measures a material's resistance to heat transfer. In Europe, the U-value is often listed; in North America, the R-value is typically used, with the primary exception being for windows. The R-value is either represented for a material of a given thickness (i.e., an R-19 fiberglass batt to fit a 2×6 stud cavity), or for a material's insulation value per inch (i.e., XPS foam is R-5/inch).

Determination of the R-value for a material is typically conducted under strictly controlled laboratory conditions: still, dry, and warm. This poses a problem, therefore, in understanding how a material performs as part of a building system. Let's look at fiberglass insulation: A batt designed to fit into a 2×6 stud wall cavity is rated at R-19, which is a measure of the conductive, internal radiant, and internal convective heat transfer. But you won't have laboratory testing conditions in your building; the presence of moisture and air flow, and lower ambient temperature — all three of which are commonly found in a stud wall in cold climates — reduces that R-value significantly. This is a very important thing to understand: R-values are *dynamic,* constantly changing in response to different conditions.[2]

The conductive performance of a wall system is not equivalent to the performance of the insulation alone. Every time insulation (say, R-19 fiberglass batt) is interrupted with a 2×6 stud (approximately R-6.8), the wall system as a whole experiences a loss of overall R-value. If you add in all the conductive losses through the framing in a standard 2×6 stud wall framed 16″ on center, the R-19 rating may plummet to R-13.[3] These conductive bypasses are called *thermal bridges;* the most frequent culprits are framing members that run through the insulation plane of the building.[4] We'll discuss this concept later in evaluating a whole-house thermal performance strategy.

Encouraging conduction isn't always a bad thing. Conduction is quite beneficial as part of certain heating strategies, such as sitting on a heated cob or stone bench or walking on a hydronically heated "radiant" floor — and wherever heat exchangers are used, such as the pipes of hot water circulating through that heated floor.

Fig. 2.6: *A "reverse thermal bridge" pattern is shown in this roof, where there are melt patterns between the vertical studs, suggesting a lack of insulation between the studs. Commonly, when insulation is present, the studs are the thermal weak link and contribute more greatly to conductive loss (thermal bridging).*
CREDIT: STEPHEN PAISLEY.

Convection

The third method of heat transfer is *convection,* the movement of heat through a fluid (which can be either a gas or a liquid). The most important thing to understand about convection is that fluid rises when heated. Whereas radiant and conductive heat move equally in all directions based on concentration gradient (barring any other variables), hot air or liquid rises through convection. This happens because as the fluid gets warmer, it expands and becomes buoyant (less dense). Being more buoyant than surrounding cooler fluid, it rises, creating stratification within a volume between upper and lower level temperatures; it also creates the motion that drives the stratification. In a building, we feel this stratification. For example, in winter, think of how much warmer it is when changing an overhead light bulb on the second story of a house, compared to lying down for a nap on the first floor; the lower floor is cooled by the drafts induced by the motion of convection. In this example, we can see a pattern: warm air rises (and exits) near the top of a building, pulling cold air into the building, further exacerbating the stratification and fueling the cycle of motion. This is called a *convective cycle* or *loop* — not so good for thermal efficiency, but quite helpful for inducing natural ventilation currents in a building.

An equation for convection is:

$$Q = h \times \Delta T \times A$$

Where Q is the heat rate, h is the coefficient of convection, ΔT is the difference in temperature between an object and its fluid environment, and A is the area of the object's surface.

This coefficient h is based on several variables, including the viscosity of the fluid and the degree of turbulence of the flow. Part of the problem with convection is that heat loss is exacerbated when temperatures are coldest — when you need heat the most. Referencing our equation, convection is partly a factor of difference in temperature across the enclosure; the colder it is outside, the greater the ΔT, the greater the convection potential, the greater the heat loss.

Infiltration and Exfiltration

To reduce convection loss, we look to a building's *air barrier* to keep air from moving through the enclosure from the exterior to the interior, and vice versa. Air barriers are designed to stop

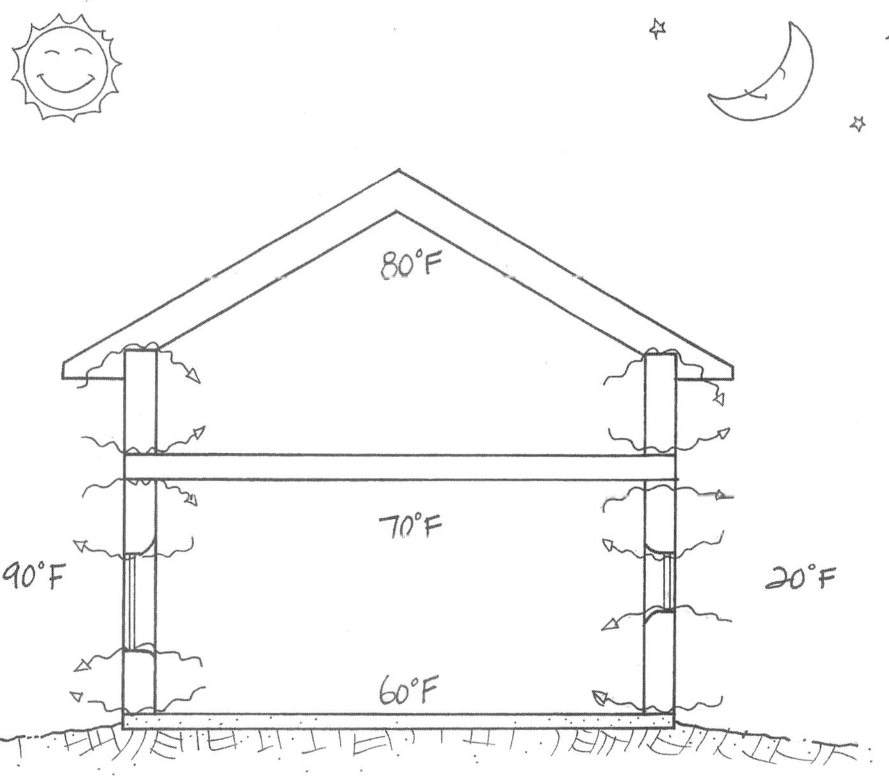

Fig. 2.7: *Air leakage into and out of building assemblies carries with it a lot of heat when the temperature difference between the inside and outside is great.*

air flow into and out of the building. This type of convection is called *infiltration* when it refers to outside air leaking into the building, and *exfiltration* when it refers to inside air leaking out through the enclosure.

Controlling convective heat flow is critical for a high-performance building. There's a saying that it is better to have a tight house that's poorly insulated than a well-insulated house that's leaky. Heat loss through convection can be extreme. A leaky house can lose a whole lot of air quickly — especially considering that convection forces are greater as temperature lowers. Whereas conductive losses are proportional to the amount of insulation that happens to be missing, convective losses can be much greater relative to the size of the hole due to the pressure dynamics at play (think of putting your thumb over a garden hose — smaller hole, bigger pressure!).

Convective losses can make occupants feel more uncomfortable, regardless of the actual quantity of heat loss. We heat our buildings to be comfortable. A draft caused by convection feels much colder than still air of the same temperature. (If we are looking to be cooled down, however, then convection is our friend.) Considering that drafts are more likely to be felt in one of the highest occupancy zones — floor level on the ground floor — even relatively minor convection heat losses can greatly reduce occupant comfort.

Interstitial Convection

There is another form of air flow that can wreak havoc on the thermal performance of a wall. In wall systems with excessive gaps and channels, convective cycles can form *within* the wall cavity; these can be created by gaps either to the

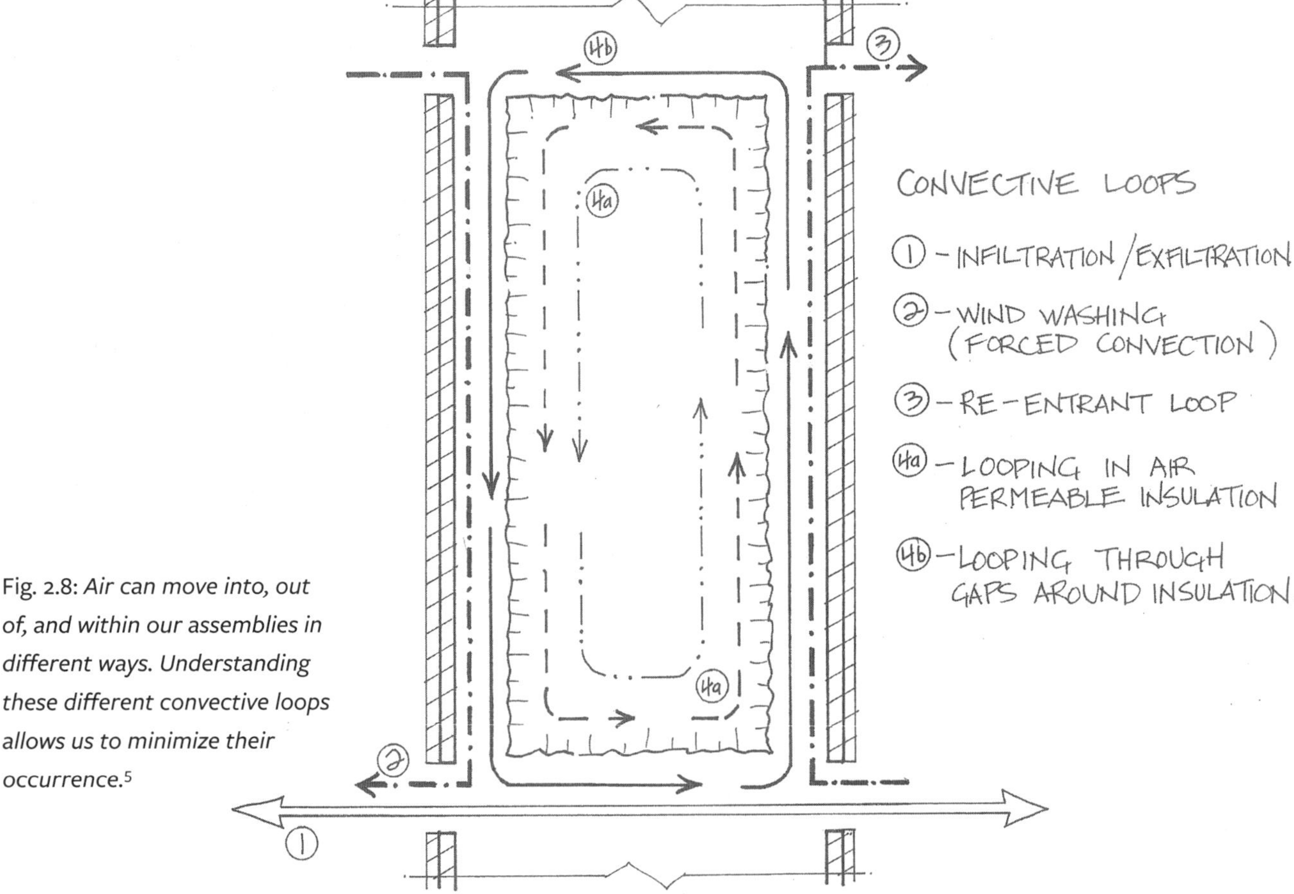

Fig. 2.8: *Air can move into, out of, and within our assemblies in different ways. Understanding these different convective loops allows us to minimize their occurrence.*[5]

interior or exterior, or both. In fact, one could have a very sound air barrier, yet still have this *interstitial convective cycling* occur within the insulation plane. Convective movement can occur in wall systems in different ways, depending on whether the air is infiltrating from the exterior or exfiltrating from the interior of the building, or if the air is looping within the cavity or through the insulation as a closed system.

This is one of the greater liabilities of batt insulation: batts are difficult to install without leaving vertical channels of air in the far corners of the cavity. These channels are perfect places for interstitial convective cycling to occur — tall, narrow, and right in the area where cold air on the outside and warm air on the inside can meet and stratify. CSG, a former EnergyStar rater in Massachusetts, has found that 5% void space in fiberglass batt installation is typical, which reduces performance from R-19 to R-11. Depending on the size and location of the gaps and the temperature differential, convective cycling can increase heat loss by 30%–50%.[6]

This same convective cycle works to our advantage when it is located in the vent space between the back of the siding and exterior face of the sheathing, as is found in a rainscreen siding system (discussed in greater detail in Chapter 6), or in the vent cavity adjacent to the roof sheathing in a vented roof, where evaporative drying is encouraged.

The Stack Effect and Wind Pressure

We've discussed the motion that comes with convection: hot air rises and exits, sucking in cold air through leaks down low in the building. The term "sucking" is a key one here, as it implies the negative pressure environment that is created in the lower half of the structure by the difference in buoyancy between indoor and outdoor air and by the increase in air pressure at lower elevations. So, in addition to a temperature gradient, we have a pressure gradient in the building — positive pressure up high, negative pressure down low (in a cold-climate winter; in a hot-climate summer in an air-conditioned

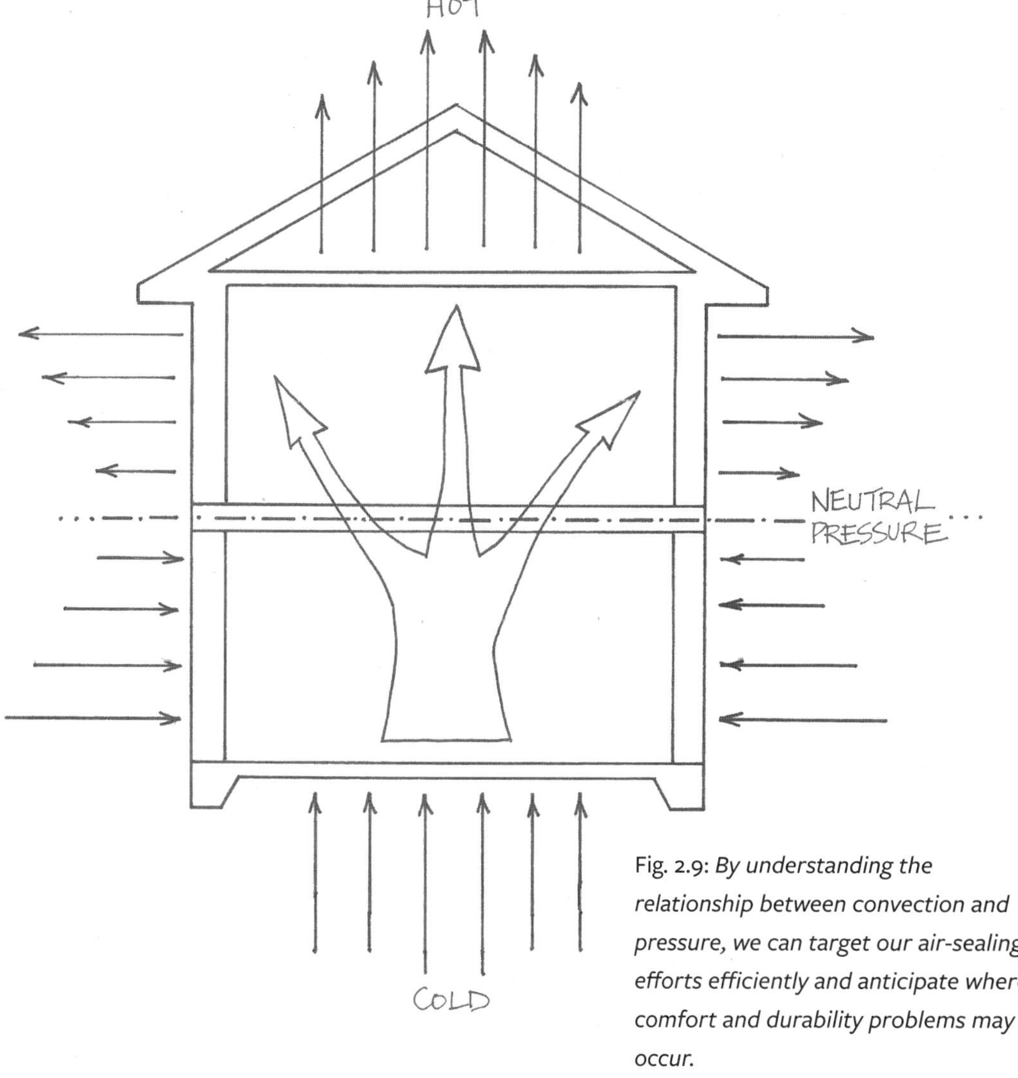

Fig. 2.9: *By understanding the relationship between convection and pressure, we can target our air-sealing efforts efficiently and anticipate where comfort and durability problems may occur.*

building, this is reversed) — and a theoretical neutral pressure plane in the volumetric center of the structure. This phenomenon is often called the *stack effect,* and is reliant on a few factors, including the temperature differential and the height of the building (taller buildings increase the effect).

If you want to cool a building, you can use this pressure dynamic to your advantage. Here's an example: at night, when the outdoor temperature is cooler than inside the building, open a window in a stuffy second-story bedroom and open a door (screened is probably preferable) downstairs. This encourages a convection loop throughout the building — the strategy is called *night flushing.* And it is the stack effect, or *chimney effect,* that allows for the smoke to flow up and out of our chimneys and not fill our houses. Unregulated, though, this pressure dynamic creates a problem. Let's look at a cold winter scenario: pressurized hot air, laden with moisture from the building, is trying to escape through every weakness and defect in the air barrier at the upstairs ceiling. Unfortunately, this is also where it is the most difficult to control that detailing because the walls and the roof often come together in complicated ways. In hot climates, the stack effect is reversed, with warm air coming into the building, being cooled by the air conditioner, dropping to the floor, and creating a positive pressure environment in the lower half of the building. Accordingly, we must pay particularly close attention to these top-of-wall and bottom-of-wall details when designing our air barriers.

Further complicating issues of pressure, we now consider the impact of pressure differentials borne of wind patterns on and about the structure. Wind blowing against a wall creates

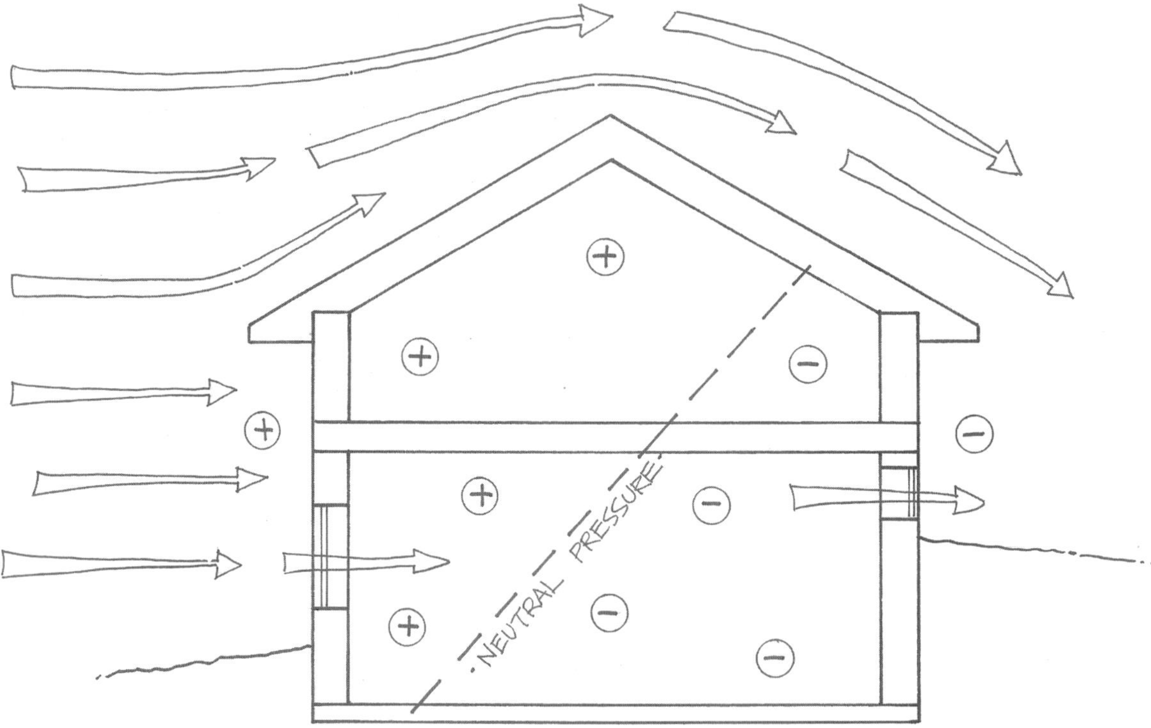

Fig. 2.10: *Wind pressures can have an even greater impact on a building than the stack effect, which may influence siting and design decisions for particularly windy sites.*

positive force on that outside wall, and a corresponding positive pressure on the interior side due to leakage. On the exterior leeward side of the building, a negative pressure environment is created. This tilts the neutral pressure plane down on the windward side, changes the internal pressure dynamics, and accelerates infiltration and exfiltration.[7]

Mechanical Convection

There are many instances in which forced convection, or mechanically powered circulation, is employed in a building. In forced convection, motion — driven by mechanical action such as a fan blowing air or a pump circulating water — is applied to the system, resulting in heat exchange. Examples of this are found in heat recovery ventilation systems (air-to-air heat exchange) and pump-circulated solar hot water systems (water-to-water heat exchange).

Latent Energy

Another form of heat transfer comes in the form of a material phase change from liquid to gas or vice versa. It takes a lot of energy to change water from liquid to gas — a pound of water will hover at the same temperature as 970 BTUs are added before changing phase to gas and rising in temperature. The energy needed for this phase change (with no temperature change) is called *the heat of vaporization,* and the stored phase-change energy in a vapor is called *latent energy.* There can be a lot of latent energy stored in the heated air in a building. With proper consideration, that energy can be contained within the building by preventing convection losses. Warm air can also be blown through a wet membrane to cool a building down with an evaporative cooling unit in a warm, dry environment.

Latent energy can be effectively captured. Many high-performance boilers and furnaces achieve improved efficiency by forcing the condensation of water vapor in the exhaust gases within a heat exchanger, thus capturing the latent heat before it is drafted out the chimney. Phase-change materials are now starting to come onto the market to take advantage of this principle; for example, drywall panels containing paraffin capsules that store latent energy by melting when ambient temperatures are warm, and release this energy when ambient temperatures cool and the wax solidifies.[8]

Thermal Mass

As opposed to insulation, which is designed to slow conductive heat flow, or an air barrier, which is designed to stop convective heat gain or loss, the *thermal mass* of a building will store heat, helping to regulate heat energy flow in two critical ways: *temperature damping* (reducing the temperature swings within a building), and *thermal lag* (delaying the transfer of heat by a period of many hours). To understand how to make use of thermal mass in a building, let's start by looking at the properties of thermal mass materials.

The first property to consider is the *specific heat capacity* of a material. This is a property that describes the amount of energy it takes to raise the temperature of that material per unit of mass. The higher the specific heat capacity, the more energy it takes to raise the temperature of the material. This is then multiplied by the density of the material to get the *volumetric heat capacity* — the "storage capacity" of that material. However, to understand the material's value as a thermal mass, we have to balance heat capacity against the *thermal conductivity* of the material, or how readily the material will distribute heat internally. When we divide the thermal conductivity (k) of the material by its density p, multiplied by its specific heat capacity c_p (which is its volumetric heat capacity, or $p \times c_p$, we arrive at its *thermal diffusivity* (a), and the equation looks like this:

$$a = k \div (p \times c_p)$$

Understanding the property of thermal diffusivity as a relationship between a material's ability to internally move heat and its ability to store heat helps us to understand the more practical effects of how we use mass to achieve temperature damping and thermal lag.

A mass wall's ability to store large amounts of heat, coupled with its moderate conductivity, allows the heat to move smoothly through the bulk of the mass and affects the way heat energy from outside is transferred to the interior climate. First, more of this energy is stored than transferred, resulting in smaller temperature swings on the interior (temperature damping). Second, the rate of transfer is delayed, and, as the interior temperatures begin to cool off at the end of the day, the energy in the wall is released into the building; the wall finishes giving off its heat and is ready to be "recharged" as the outdoor temperatures rise again at the beginning of the day. This time factor is of particular benefit when the mass is used in such a way to take advantage of the "diurnal swing" (temperatures rising and falling over a 24-hour period), providing stored heat energy to the building when the sun is down, and storing excess heat energy when the sun is out.[9]

Applications for thermal mass vary depending on climate and heat source. In climates

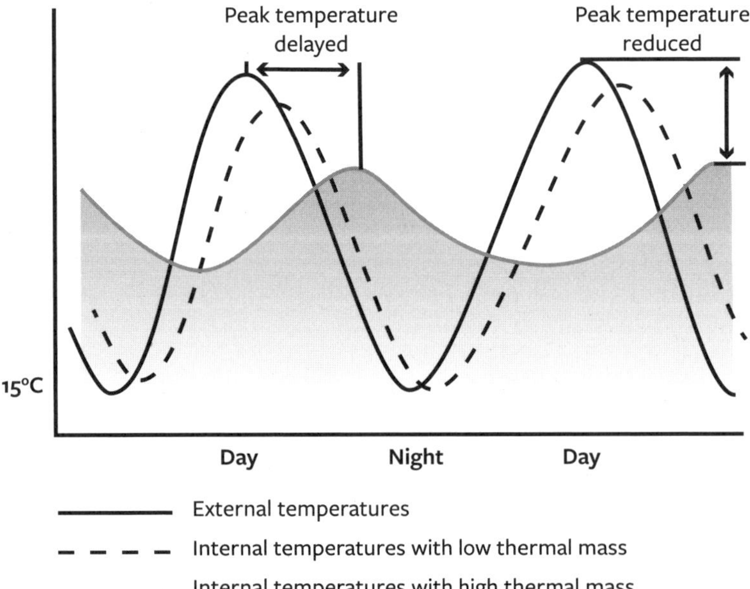

Fig. 2.11: *Mass has the effect of both damping the extremity of temperature swing and delaying the cycle, which can elegantly time in sync with the diurnal cycle, thus maximizing energy benefits.*
SOURCE: BLUE RIDGE ICF AND CONCRETECENTRE.COM

where the diurnal swing reliably provides nighttime temperatures below and daytime temperatures above the indoor comfort temperature (such as the desert southwestern United States), enclosure walls of primarily massive, non-insulative materials can effectively control the interior temperature, as excess heat from the daytime sun is stored, keeping interior conditions cool, and then radiated into the building at night when it is needed. However, in locations where it is consistently hotter than comfort temperatures at night (such as the hot, humid regions of the southeastern United States), the cooling potential at night is reduced, and the house effectively becomes a low-temperature oven. Likewise, in regions where it is consistently cooler outside during the day than the desired indoor temperature (much of the northern half of North America), there is a steady loss of heat to the outdoors, resulting in poor performance of the building.[10]

A good solution for the use of mass in climates where insulation is required in the enclosure is to place the mass inside the building. In the case of passive solar design, the mass should be positioned in direct exposure to incoming sunlight. Mass can also be heated by other means, such as circulating tubes of fluid (such as in a hydronic, or "radiant" floor) or encasing a combustion heater (such as in a wood mass heater). In these cases, the same effects of charging up the mass with thermal energy

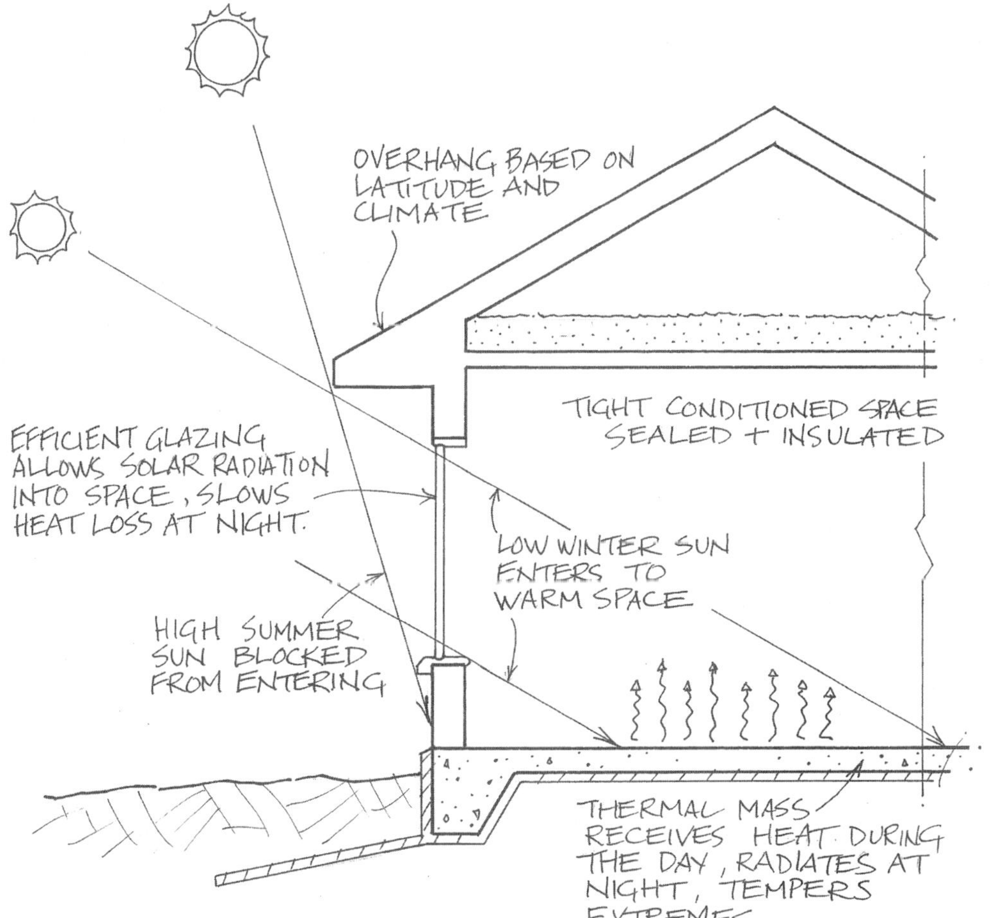

Fig. 2.12: *Thermal mass can be employed in passive solar design strategies in climates where high-mass walls are not practical.*

to slowly radiate into the building when heat is needed are realized, but are contained within a good thermal enclosure, allowing for efficiencies in both the production and distribution of heat.[11]

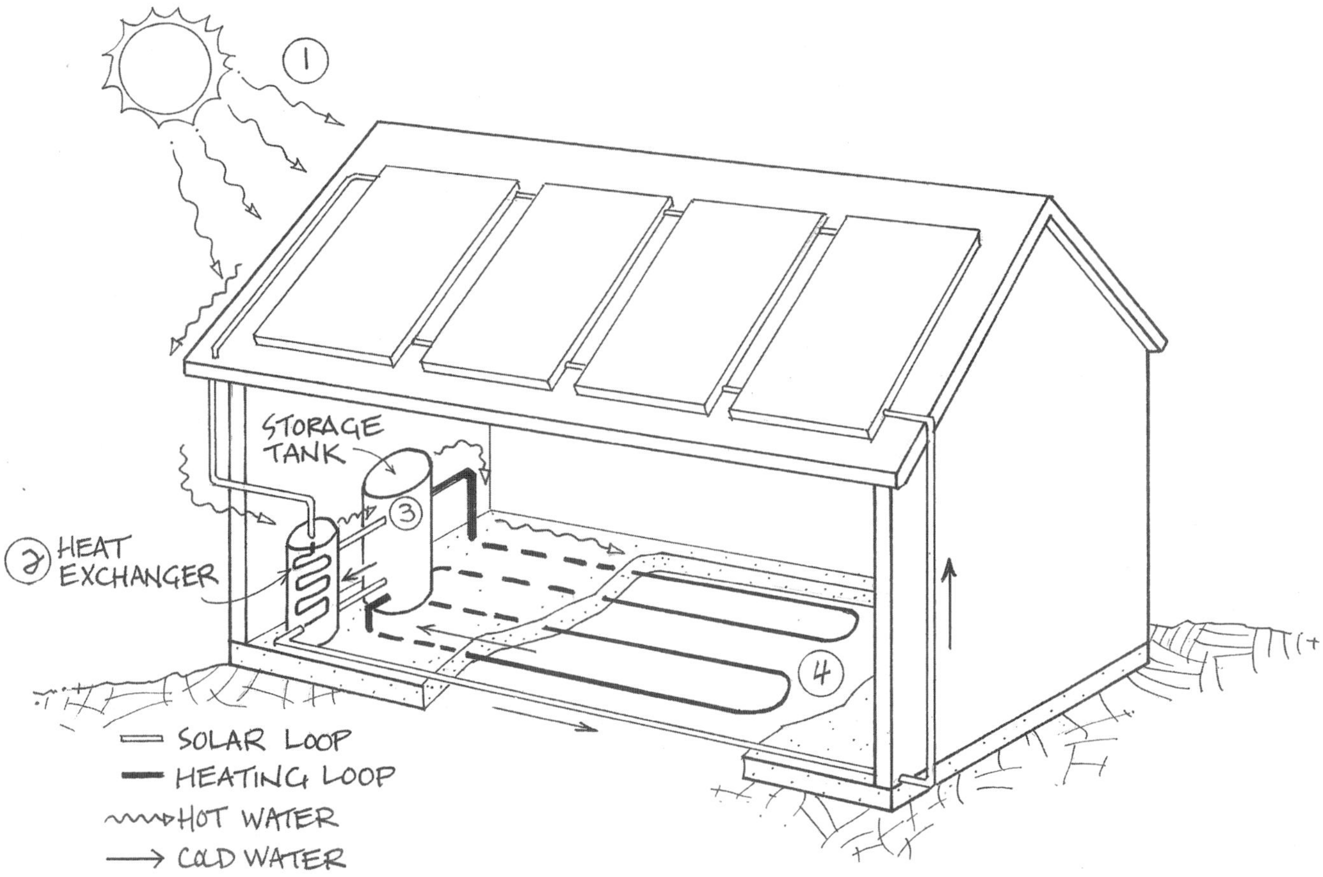

Fig. 2.13: *A solar hot water system features all forms of heat transfer: 1) solar radiation warms the collector fluid, which 2) conducts to the tank fluid, which 3) rises through convection for distribution, and can be 4) banked in the thermal mass of an earthen floor.*

Chapter 3

Moisture and Hygrothermal Dynamics

WATER IS AN AMAZING, UNIQUE SUBSTANCE that is the foundation of biological life on our planet. It is also the most destructive element we face in creating durable shelters. This is as it should be; as stated in the second law of thermodynamics, all complex systems in the universe — people, ecological systems, buildings — are driven by entropy toward disorder and breakdown. Water is simply doing its job as a primary vehicle through which entropy can do its work. For us to be able to build responsibly, we must design for water from the beginning — not leave it as an afterthought or secondary consideration. And to do that, we must understand what water is and how it moves.

Hygrodynamics 101

The molecular structure of water is one of the most commonly known: H_2O, two hydrogen atoms bonded to a single oxygen atom. This molecular structure is very important in understanding one of water's most important characteristics. Because the positively charged hydrogen atoms are weighted to one side of the negatively charged oxygen atom, the molecule acts as a tiny electromagnet — it is *polar,* permanently electrically charged — ready, willing, and able to bond with a host of substances: your skin, a wall, other water molecules. This "charged" property is of great consequence to the understanding of how water interacts with materials in a building.

Water is the only substance that exists in all three major phases in the natural conditions of our daily lives (except for those of you living in climate zone 1). We call it "ice" or "snow" in solid phase, "water" in liquid phase, and "vapor" in gaseous phase. When liquid water is heated, the molecules are excited, and the hydrogen bonds between the molecules break, forming smaller and smaller clumps of water molecules. When enough heat is added, the individual molecules begin to separate, changing phase into vapor molecules, suspended in the air. Phase change in this direction is *endothermic,* requiring heat. It takes an incredible amount of energy to break those strong hydrogen bonds; note that the temperature of water does not change during this phase-change process.

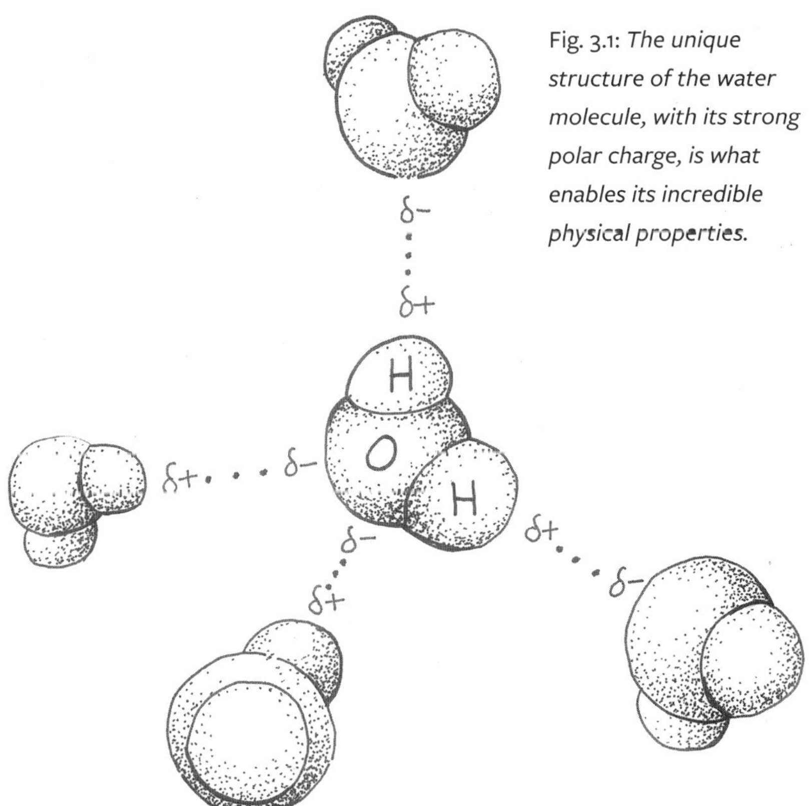

Fig. 3.1: *The unique structure of the water molecule, with its strong polar charge, is what enables its incredible physical properties.*

Moving in the other direction, as water is cooled and heat leaves the system, the molecules cluster more closely together and arrange in a variety of different ways to create the many different forms of snow and ice in their solid phase; the process is *exothermic,* giving off heat, and explains why condensing fog melts snow so effectively. When the foggy, humid air hits the cold surface of the snow, the air temperature cools, causing the water vapor in the air to condense, releasing large amounts of latent energy as it changes phase from gas to liquid; this energy release melts the snow very quickly.

When water moves from liquid to gaseous form, it is called *evaporation*; moving from gas to liquid, we call it *condensation.* Liquid-to-solid phase change is called *freezing,* whereas the reverse is called *thawing.* Moving from gaseous form straight to solid is called *frosting,* or *deposition.* When water moves from solid directly to gaseous form, it is called *sublimation.*

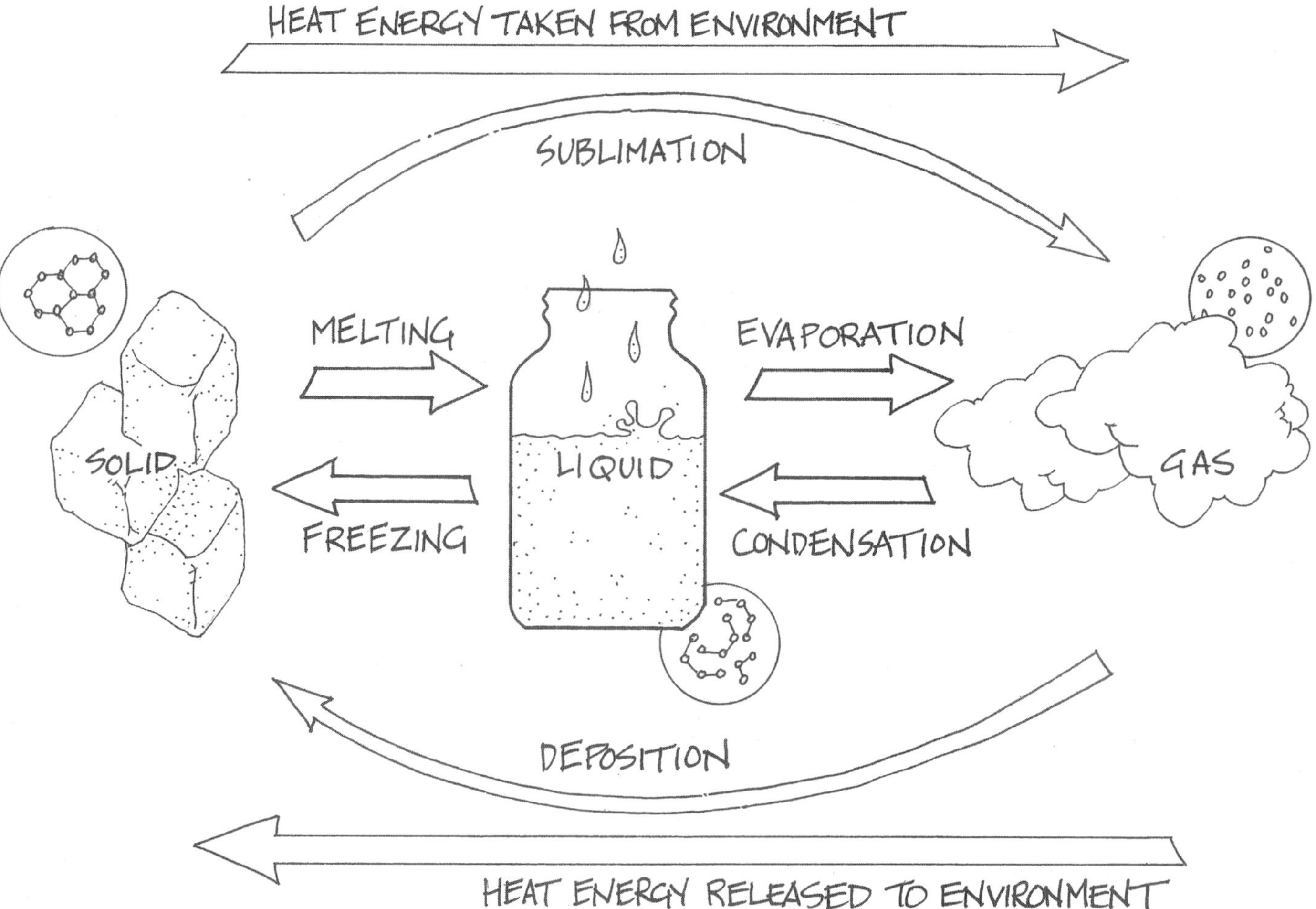

Fig. 3.2: *As water changes phase and undergoes physical transformation, large amounts of energy are absorbed or released as the powerful hydrogen bonds between molecules are created or destroyed.*

Water and Materials

There are many ways in which water interacts with building materials, and understanding some key terminology will help in the discussion of strategies:

Diffusion:

The movement of water through a material is called *diffusion.* Water moves into and through materials in two main ways: vapor diffusion and liquid diffusion.

Adsorption and absorption:

When polar water molecules bond with the surface of a material as a film, the process is called *adsorption.* When bulk water enters and fills the pores of a solid material, the process is called *absorption.* The difference between the two mechanisms is between a thin film and a more multi-layered volume of water, but the moisture can enter as either a vapor or a liquid.

Saturation:

Once all of the pore spaces have been filled, the material is said to be *saturated.* Absorption can add significant moisture to a material; in the case of wood, upward of 25%–30% moisture content by weight can be realized through absorption into the cells at 98% RH (relative humidity); this swells the cell walls, causing the wood to expand in dimension. As water molecules leave the cells, the wood shrinks, potentially causing twisting, splitting, and cupping of the wood. This, on a molecular level, is why a wood door may stick in a humid season, and why a large wooden timber may check and twist as it dries.

Vapor pressure and vapor drive:

The rates and directions of all vapor-form moisture transfer are based on a few main factors:

- concentration gradient (moves from high to low absolute humidity; this is related to vapor pressure).

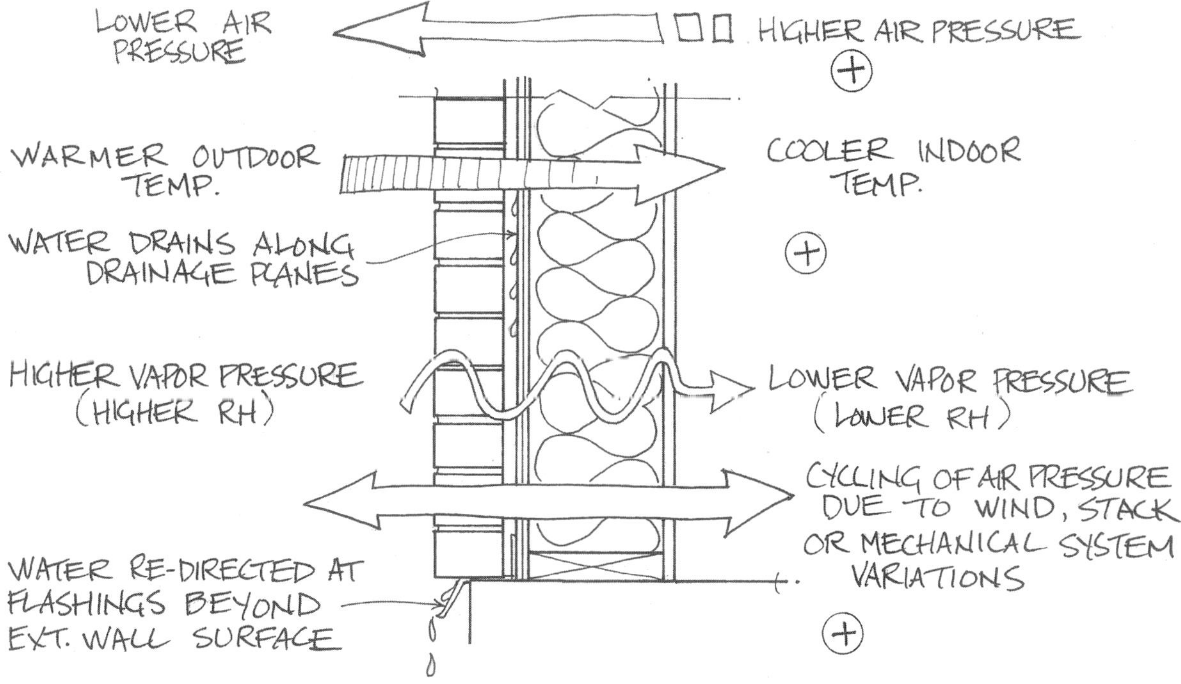

Fig. 3.3: *Vapor drive is governed by a few factors, including relative humidity, temperature, and air pressure. Accordingly, the vapor drive changes in both direction and intensity over the course of any given day. Often, relative humidity and temperature will induce a drive in one direction, overriding relative air pressure pushing in the opposing direction, as is pictured here (higher relative humidity and temperature outdoors, higher air pressure indoors).*

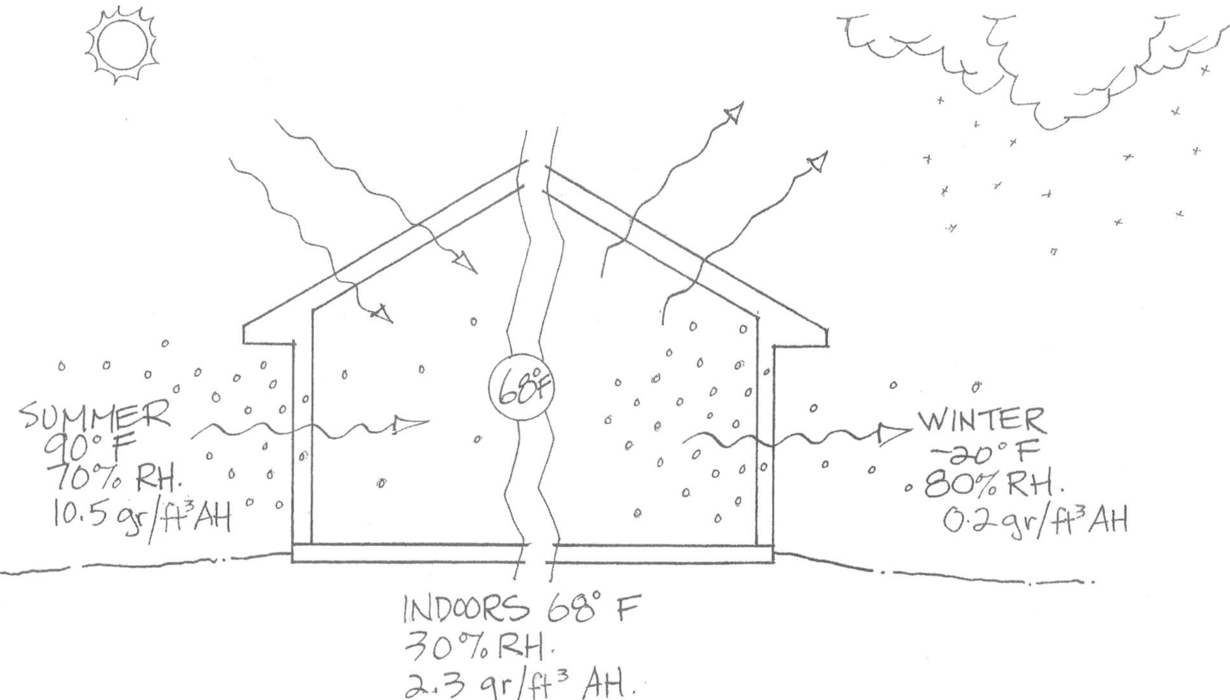

Fig. 3.4: *Vapor pressure refers to the concentration gradient of absolute humidity (AH, shown here in grains per cubic foot, or gr/ft³), and is a factor of vapor drive. Vapor diffusion moves from high vapor pressure to low vapor pressure. Notice that even though the winter relative humidity (RH) is high given the low temperature, the absolute humidity is low, resulting in vapor diffusion to the outdoors.*

- air pressure gradient (moves from high to low air pressure, and not the same thing as vapor pressure).
- temperature gradient (moves from high to low temperature).

Combined, these factors create a *vapor drive,* forcing water vapor molecules to diffuse through the materials of a building enclosure in the direction of the vapor drive. Generally speaking, the direction of the vapor drive is to the outside in a cold winter, and to the inside during a hot summer (especially when air conditioning is operating in the building).

Vapor control layers, retarders, and barriers: In our buildings, we use *vapor control layers,* often called *vapor retarders,* to control the diffusion of moisture through the enclosure materials by way of vapor drive. In below-grade applications (such as basement walls or slabs), this drive is always to the inside, as it is more humid in the earth than the desired condition inside the building. In this case, a *vapor barrier,* which is a vapor retarder that stops practically all vapor diffusion, is appropriate. In above-grade applications (walls and ceilings/roofs), the drive may vary, from region to region, season to season, day to day, and even hour to hour depending on the exterior and interior climate conditions. Therefore, anticipating this drive can be difficult, and strategies that take into consideration the dynamic nature of the vapor drive in our buildings are important, as we will discuss in Chapter 4.[1]

Permeability and permeance:
We can choose vapor retarders that have varying degrees of *permeability,* which is the property that describes the propensity of a material to allow vapor diffusion at a given temperature, relative humidity, and atmospheric pressure, independent of its thickness. *Permeance* is the actual amount of water vapor that diffuses; and this depends on the material's thickness. Some vapor retarders and materials are *vapor-variable,* meaning their permeability will change as the relative humidity changes. The permeability of our vapor control layers can vary depending on the nature of their design in response to the conditions of the assembly.

One of the biggest misunderstandings in how our buildings work is the confusion surrounding the movement of vapor into our walls and roofs. So far, we have been discussing vapor diffusing into the materials of the assembly through vapor drive. However, far greater volumes of vapor can be carried into our assemblies by bulk air leakage through convection. As illustrated in Figure 3.5, the leakage of air creates more transfer of vapor than diffusion by many orders of magnitude, and it is the air barrier that is responsible for controlling transfer along this pathway. This is a critically important concept that we must explore in the context of how heat and moisture move together.[2]

The leakage of air creates more transfer of vapor than diffusion by many orders of magnitude, and it is the air barrier that is responsible for controlling transfer along this pathway.

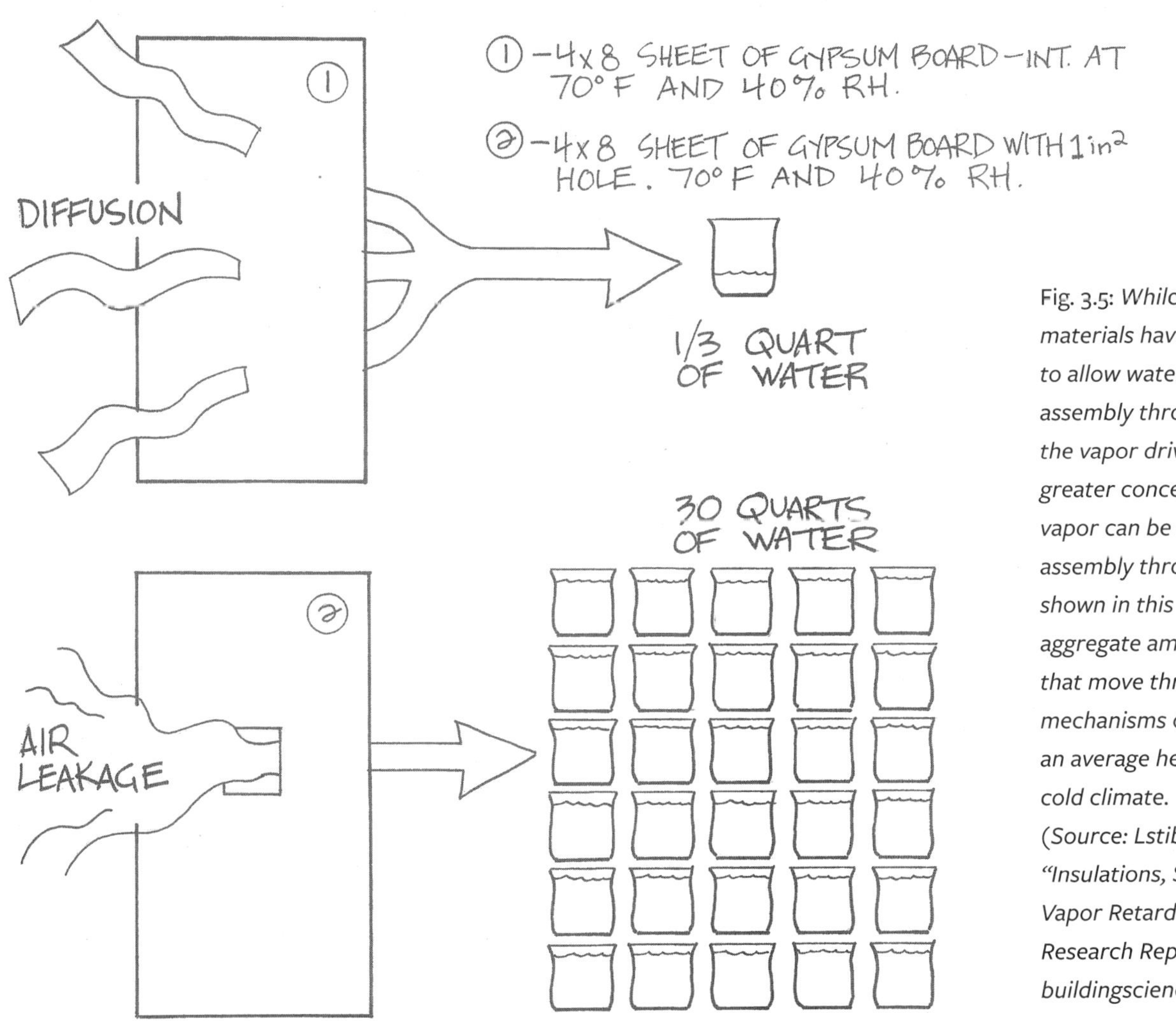

Fig. 3.5: *While vapor permeable materials have the potential to allow water vapor into an assembly through diffusion when the vapor drive is sufficient, far greater concentrations of water vapor can be carried into the assembly through air infiltration; shown in this illustration is the aggregate amount of moisture that move through these two mechanisms over the course of an average heating season in a cold climate. (Source: Lstiburek, Joe: "Insulations, Sheathings, and Vapor Retarders," Research Report 0412, 11/04, buildingscience.com)*

Hygrothermal Dynamics

Already, in our exploration of how vapor moves through building enclosures, we see the relationship of water and heat, as vapor drive is partially influenced by temperature, and convective heat transfer is the prime mechanism for vapor transfer as well. Putting these two systems together is fundamental in understanding how our buildings work, troubleshooting problems as they occur, and identifying solutions that will avoid problems in the first place.

The mass of vapor molecules in a given volume of air independent of temperature is called the *absolute humidity,* expressed as a percentage. This number is helpful for determining vapor pressure. However, this is not a great metric for how we experience and understand the humidity of the air around us. Rather, we tend to think in terms of *relative humidity* (RH), or the mass of vapor molecules in a given volume of air at a given temperature relative to the air's total carrying capacity at *saturation* expressed as a percentage. As we try to exceed 100% RH, we get condensation (rain, for example, or water on window glass). As the temperature is raised, a unit of air with a certain amount of moisture becomes less dense, and can hold more water vapor — resulting in a decrease in RH, as the amount of water in the air is a smaller portion of the increased capacity. As the air becomes cooler, it becomes more dense; the molecules crowd closer together, and there is less carrying capacity for water vapor — resulting in an increase of RH. Note that there isn't more water being added — the absolute humidity remains the same — but the container (carrying capacity of the air) is getting smaller as the temperature drops. At some point, the capacity will get so low that the water vapor molecules will condense into liquid. This temperature is called the *dew point;* the dew point is relative to the absolute humidity of the air and the temperature at which the relative humidity reaches 100%. This can happen on a daily cycle outdoors; it may also be happening in your wall or roof assemblies.

In a cold climate in the winter, the vapor drive is predominately from the warmer, moister inside to the colder, drier outside. The convective

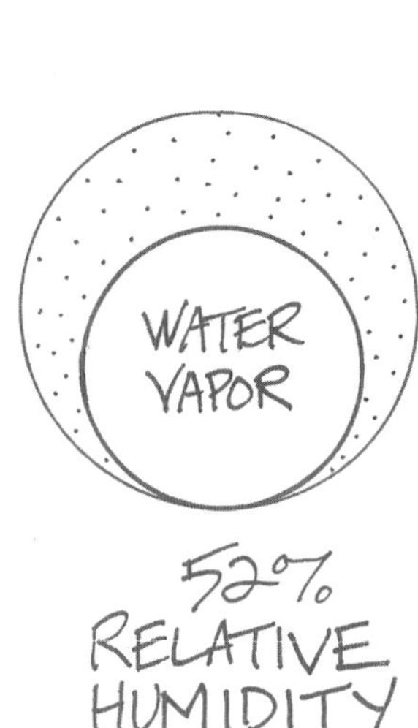

Fig. 3.6: *As air temperature drops, the air gets denser, and the relative humidity increases, even though no additional vapor has been introduced into the system.*

pressure is also strong in the winter when the difference in temperature between the inside and outside is greatest, and the positive pressure inside the building, pushing out, increases the higher up into the building you go (on a still day). Therefore, the force of convection has the ability to move huge amounts of water vapor through imperfections in the building's air barrier into the building enclosure, far more than the potential of diffusion alone. If this vapor-rich warm air hits a cold surface — such as the uninsulated exterior sheathing of a building — and the air cools to the dew point, condensation will occur on this surface, and damage may occur if the assembly's storage and drying capacity isn't sufficient (this will be discussed further in Chapter 4).[3]

In hot-humid climates where the building is air conditioned, the same can happen in reverse, with hot humid air rushing in through inward convection, and humidity moving into the assembly through inward vapor drive loading the assembly with vapor, which can build up and condense on the interior side as it cools.

Creating a solid air barrier is not just a thermal strategy, it is a key durability strategy as well. Accordingly, the convective currents that exist within a building, or within an assembly, will exacerbate the transport of humidity; understanding these cycles can help point toward

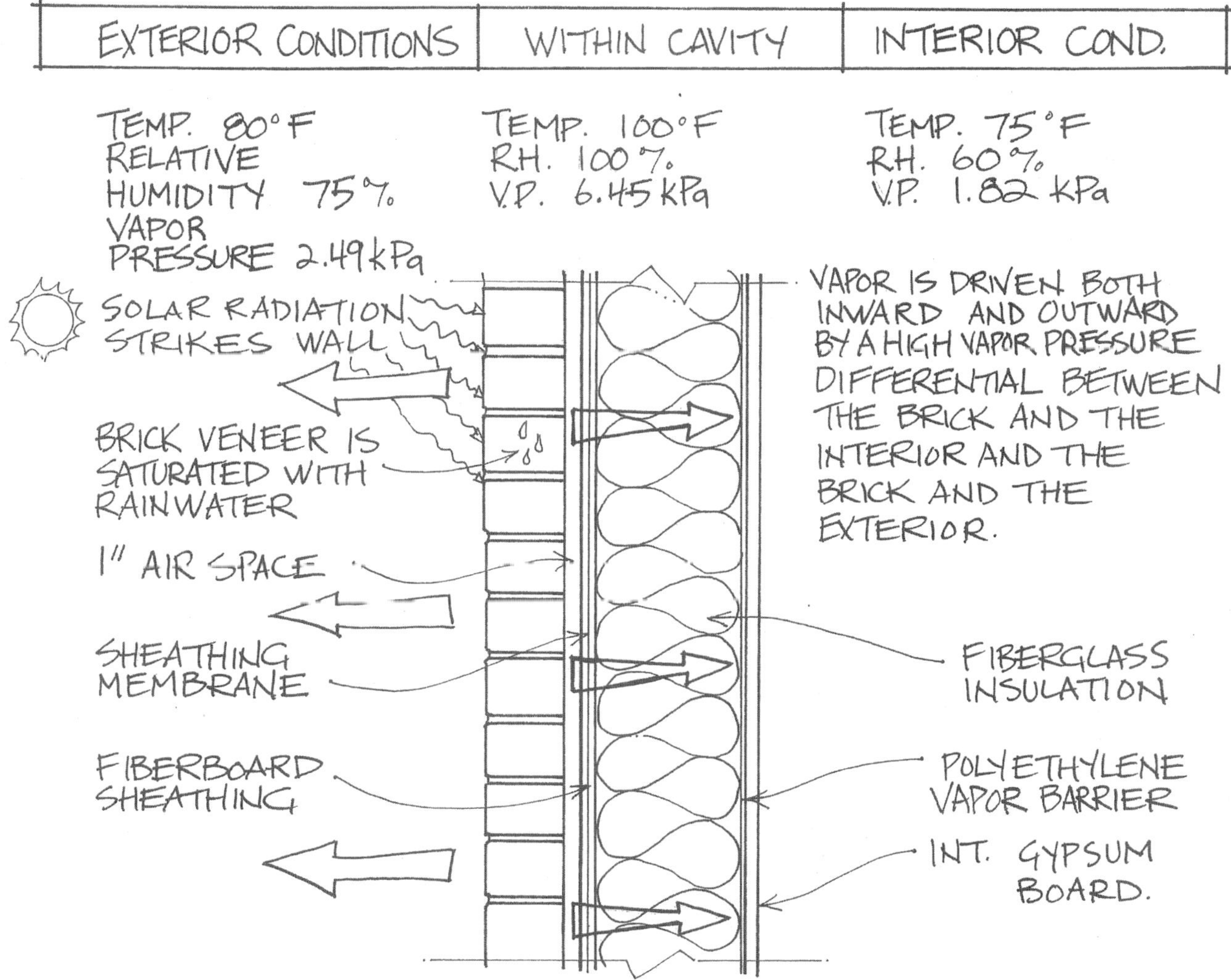

Fig. 3.7: *The phenomenon of solar-powered vapor drive can wreak havoc on walls featuring both reservoir cladding and interior vapor barriers.*

strategies that avoid moisture buildup within our buildings.[4]

This may all seem straightforward, but it can actually be quite tricky because often different things will happen at the same time. Let's look at a south-facing earthen or brick wall on a sunny day, just after a rainstorm has soaked the wall. Rainwater has been absorbed into the outer layer of the wall. The sun is doing two things to the water: 1) it is evaporating water on the surface, returning it to the atmosphere, and 2) it is driving the water deeper into the wall. Why? Take a look at the vapor pressure (concentration gradient) in Figure 3.7. There's 2.49 kPa (kilopascals, a measurement of vapor pressure) outside — due to the high concentration of vapor in the air — and 1.82 kPa inside. Take a look at the temperature gradient: 80°F (26.6°C) outside, 75°F (23.8°C) inside. In the cavity space behind the brick, the temperature and vapor pressure are even higher than the outdoor environment. Where's the vapor going to drive? Even if the building has some positive pressure, vapor pressure and thermally driven diffusion will overcome opposing forces of air pressure, and there will be an inward vapor drive. This is a basic example of how unintuitive water's movements can be; so it is important to have a thorough understanding of how and why it moves the way it does.[5]

In the days before insulation was common, our old buildings in cold climates did not have condensation problems — despite the fact that they were incredibly drafty. The reason is that the tremendous volume and rate of heat loss would help to drive the moisture out of the assembly. Inefficient, to be sure, but it certainly helped with durability! If large amounts of insulation are added to a roof or walls, as is common in energy retrofits, the amount of heat entering the enclosure is dramatically reduced. While this is certainly the intended goal from an efficiency standpoint, it is important to understand that there is less energy in the assembly to compel drying, and the components of the assembly on the "cold side" (such as uninsulated exterior sheathing in cold climates, or interior drywall in an air-conditioned room in a hot-humid climate) will be colder, and therefore induce greater amounts of condensation (when air within the assembly cools and reaches the dew point). From what we know about the potential for convective losses to carry large quantities of vapor into an enclosure, the importance of coupling a high-quality air barrier with the insulation layer becomes all the more critical as the insulation layer becomes more robust.[6]

This is not to suggest that, from a building science perspective, we should all revert back to the days of leaving our buildings uninsulated for the sake of avoiding condensation in our walls. Rather, it is to say that as we continue to make our buildings more thermally efficient, we must also ensure that they become more durable in response to their decreased drying potential. To accomplish this successfully, we need to employ strategies for moisture control; the right approach will avoid the greatest risks while enhancing the greatest benefits.

Part III:

Developing Strategies

Credit: Ace McArleton — New Frameworks Natural Design/Build.

Chapter 4

Moisture Control

We can organize the parts of our enclosures into four different control layers that do four different things:

1) the *water control layer,* often known as a *weather-resistant barrier* (WRB), connects with flashing and keeps water out.
2) the *air control layer,* or *air barrier,* limits air flow into and out of building assemblies.
3) the *thermal control layer* (in modern buildings, primarily *insulation*) slows the movement of heat in or out.
4) the *vapor control layer,* or *vapor retarder,* limits vapor movement via diffusion through our building assemblies.

We begin our exploration of the enclosure in this chapter, as we walk through the different moisture pressures our enclosures face and the role each control layer plays in managing moisture.

To start, we should identify why this deep analysis of moisture is merited. Water's entry into buildings, when not adequately controlled,

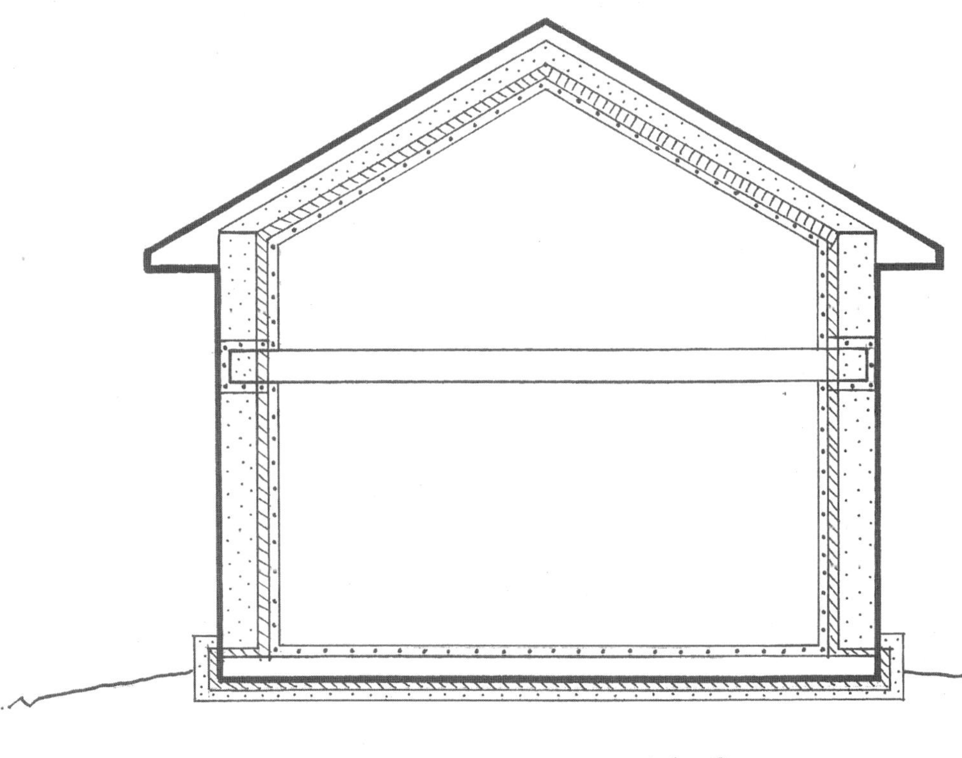

Fig. 4.1: *A well-designed enclosure does four things:*
1) keeps water out;
2) controls air flow into and out of the building;
3) keeps heat in or out as desired;
4) controls/allows vapor migration. Sometimes one material will do more than one job, but each job must be done.

causes many different problems. The absorption of liquid water can cause materials such as wood and clay to swell (pop) and then shrink (split), which in turn can open up spaces for additional water and/or air leakage. In masonry, we see water as a solvent carrying dissolved salts to the surface of the wall, where they are deposited as a white film (known as *efflorescence,*) after the water evaporates; this condition can be merely cosmetic, or it can lead to spalling, or even outright crumbling due to the enormous osmotic pressures of water. Water's polar nature also causes destructive chemical reactions with the iron present in many metals, leading to oxidation and decay of materials.

Perhaps most notorious and nefarious to us in the building world is water's role as a key ingredient in micro-biological habitat (see

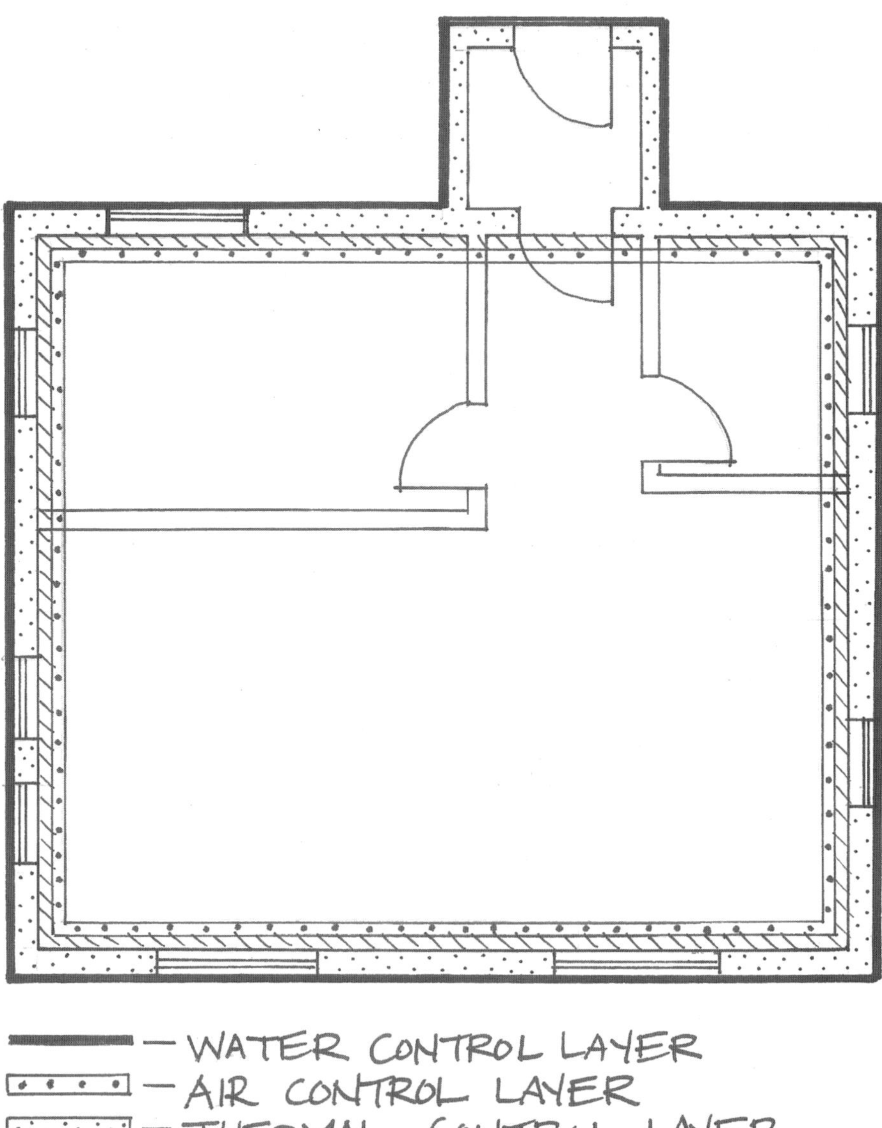

Fig. 4.2: *The layers of the enclosure must be well-defined and continuous to be effective. The marks of a high-performance enclosure are well-designed and well-executed transitions and details!*

The Six Ingredients for Mold

There are hundreds of types of mold, and buildings are riddled with uncountable numbers of their spores. So what keeps us from being overrun by blooms in our bedrooms? For a mold bloom to occur, six conditions all must be present; if any one of them is missing, mold cannot grow (note: we are referring to common molds found in most buildings, there are exceptions to any of these conditions)[1]:

- Food — Mold needs an organic material to eat, and it prefers cellulosic material, like straw, paper, and wood.
- Oxygen — Molds need oxygen, even small amounts, to grow. Unfortunately, your walls aren't going to be tight enough to create anaerobic (oxygen-free) conditions, despite your best air-sealing efforts.
- Temperature — Most molds like to be within a temperature range of 40–110°F (4.5 to 43.5°C) to grow, easily encompassing the standard temperature range of our homes.
- Darkness — UV light is no good for most molds; but if the sun is shining inside your walls, you have more substantial problems to address.
- Moisture — Mold growth can start as early as 70% RH (attainable through vapor diffusion alone), although more substantial decay doesn't begin until much higher RH numbers are reached (approximately 97%), generally in the presence of liquid moisture.[2]
- pH — Molds generally like a pH range between 3 and 7. Lime has for centuries been used as a hygienic finish to treat mold by virtue of its high alkalinity (high pH).

sidebar: "The Six Ingredients for Mold"). Two big problems can arise from mold and mildew (both are types of fungus); for one, they can have noxious effects on occupants, from unpleasant odors, to mild allergic reactions, to severe respiratory distress or other symptoms in extreme cases or for sensitive individuals. For another, they can cause rot and decay in our building materials, to the point of condemnation. While the other issues raised in this section are quite serious, mold growth and decay are the most feared of any that water inflicts on our structures. Insect infestation marks another biologic symptom of moisture. Insects are indicators of moisture problems, and can be thought of as bioindicators for a deeper problem in need of addressing; carpenter ants, termites, and other burrowing bugs cause additional damage to materials.

Fig. 4.3: *The six ingredients for mold were clearly present in this interior, as evidenced by the mold blooming on the ceiling.*
CREDIT: ACE MCARLETON — NEW FRAMEWORKS NATURAL DESIGN/BUILD.

Moisture Control: The 5 D's

It is difficult — and in some ways dangerously myopic — to look at moisture management only on an assembly-by-assembly basis, as it is a systems approach to the building that is required to deal with a systems-based natural phenomenon. To that end, we borrow a page from building scientist Dr. John Straube, who, in the book *Design for Straw Bale Buildings,* contributes a valuable strategy for handling moisture in buildings based not only on keeping them from getting wet, but also by incorporating principles of drying and safe storage within the assembly as well; his is a balanced approach that works with the natural patterns of moisture movement in our buildings. Here are the "Five D's" of moisture control, in rough order of priority:

Design

As Straube himself says: "The best moisture control strategies always involve designing problems OUT — not solving them after they have been needlessly designed into the enclosure."[3] Troubleshooting moisture problems is rarely as effective and never as affordable as avoiding them in the first place through good design!

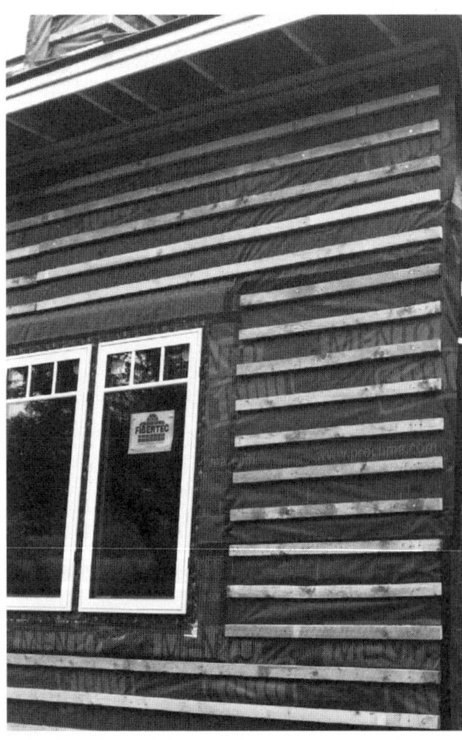

Fig. 4.4: *Providing drainage potential behind siding in wet climates is a terrific way to help ensure the durability of the wall; note the horizontal strapping has drainage channels routed behind to allow for full wall drainage. Deflection is provided by the weather-resistant barrier (WRB), which is also acting as an air barrier (shown), as well as the cladding to be installed (not shown).*
CREDIT: ACE MCARLETON — NEW FRAMEWORKS NATURAL DESIGN/BUILD.

Deflection

The greatest moisture threat is bulk wetting, primarily from rain and snow but also from surface and groundwater. Deflecting this water away from our buildings is therefore the first place to start, and this is the job we ask first of our roofs and overhangs, and then of our water control layer. Further, we ask our air control layer to deflect vapor-laden air from entering our assemblies, either from the inside during the winter or from the outside during the summer.

Drainage

Drainage strategies rely on providing water an easy path away from or out of the assembly once it has made its way through the "first line of defense." This requires gutters, downspouts and leaders (at least 10 feet from the foundation), proper site slope (a minimum of ½" per foot for ten feet), and good foundation drains (ideally, to daylight). Secondary strategies include exterior façade elements such as appropriate flashings, sloped window sills, rainscreens, and window pans. Many roofs are also designed to allow water that makes its way through the primary roofing to safely drain out of the assembly over a well-flashed underlayment. Given the risk of water entry into our buildings, particularly in wet climates, incorporating drainage details into our water control layers is well worth the investment of time, energy, and materials.

Deposit (Storage)

If we assume that we will not be able to execute the detailing of our assemblies to perfection and that some wetting will occur — whether from leaks or condensation — there is a great benefit in using materials that safely store an appreciable amount of water for enough time to allow for drying. This concept requires us to evaluate the hygric properties of our building materials, as described in Chapter 3. Hydrophilic materials that exhibit good durability, such as wood, cellulose, clay, and lime, can provide a "moisture battery" in our walls and roofs to deal with incidental wetting events. However, there are a few caveats here that are very important to address: storage without drying will lead to building failure — if hydrophilic materials inside a wall get wetter faster than they can dry out, the storage capacity will only serve to keep water in the assembly and will be a liability rather than an asset. We should also note that there is an opposing strategy that favors the use of hydrophobic materials, such as foam, steel, and high-density/sealed masonry, with the idea that water should be encouraged to leave our building assemblies as quickly as possible under all conditions, and not encouraged to be held for any period of time. Plenty of assemblies built with this approach are highly durable, and this strategy is particularly relevant in high-moisture environments and flood-prone sites, where the amount of wetting may overwhelm any storage potential.

Drying

The ability for moisture inside our assemblies to safely dry to the inside or outside — whether or not storage is incorporated — is a critical feature in all but a small handful of assemblies. To provide for drying, first we must understand the direction of the dominant vapor drive for any given season in our climate. We can then select a strategy that encourages drying either to the inside, to the outside, or to both sides. Governing this is the permeability of the materials in our assembly. Vapor-permeable materials must be specified where drying is desired, whereas materials with less permeability (either semi-impermeable vapor retarders or impermeable vapor barriers, if required) can be used to limit or stop vapor diffusion. Developing a strategy for the drying potential of our building assemblies is a critical component of the design phase for moisture control.

Fig. 4.5: *Bulk wetting from poorly flashed windows is causing this sheathing to fail. The drying potential cannot keep up with the moisture load, and the sheathing cannot safely store that much water — especially OSB (pictured), which is more prone to rot when saturated than plywood or solid wood.*
Credit: Maria Klemperer-Johnson — Hammerstone School.

Fig. 4.6: *Vapor control layers limit drying, as well as limiting inward vapor diffusion. Vapor-variable membranes offer the best of both: vapor-closed when RH is in normal range, but will become vapor-open to promote drying when RH in the cavity becomes elevated.*
Credit: Ace McArleton — New Frameworks Natural Design/Build.

Sources of Moisture

Before we can fully develop an approach, we need to first understand how moisture is entering our buildings. This can be much easier said than done when trying to troubleshoot a leak; however, getting clear about the source is necessary to effectively deal with the problem at its root. Following is a basic survey of the categories of moisture entry into our buildings.

Rain and Snow

By far, the most significant source of moisture damage in buildings in North America comes from the bulk-loading of liquid water from rain and snow. This is the first place to start addressing water control — don't bother worrying about vapor and condensation if you haven't thought through your water control layer details! Dr. Joe Lstiburek of the Building Science Corporation offers a simple and effective strategy for handling rain water control, in order of priority[4]:

- Drain the site
- Drain the building
- Drain the assembly
- Drain the opening
- Drain the component
- Drain the material
- Summary: Drain everything!

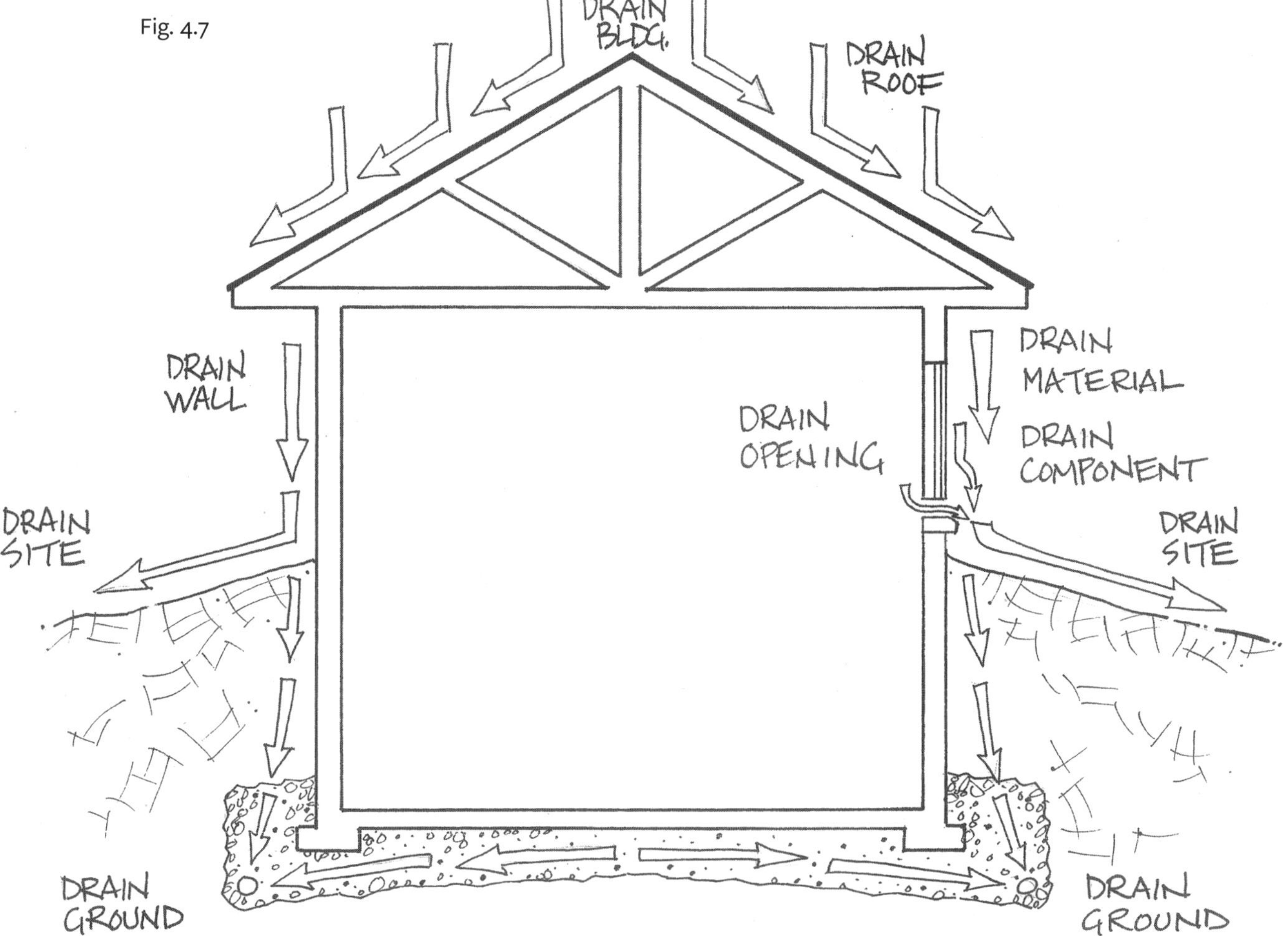

Fig. 4.7

Driving Rain

The extent of risk — and by association, the required detailing — is proportional to the amount of rain and wind exposure on a site. Climate and microclimate are the biggest factors, and siting a building with this in mind may reduce exposure, as may adjusting a building's height, shape, or profile to minimize surface exposure to rain (designing large overhangs can also help), particularly on sides of the building facing a dominant wind. Gravity is only one influence on rainfall; wind pressures (either directly driven or by creating a differential pressure space in the assembly, such as a negative pressure space created behind cladding) can drive rain in unpredictable ways and through seemingly protected barriers (i.e., through small gaps in siding or trim). Once flowing on the surface of a building, surface tension can cause water to move upside-down and backward (i.e., along the bottom of a window sill), while capillarity can cause water to be sucked into porous materials (i.e., wood or plaster), small cracks, or the overlap between clapboards.

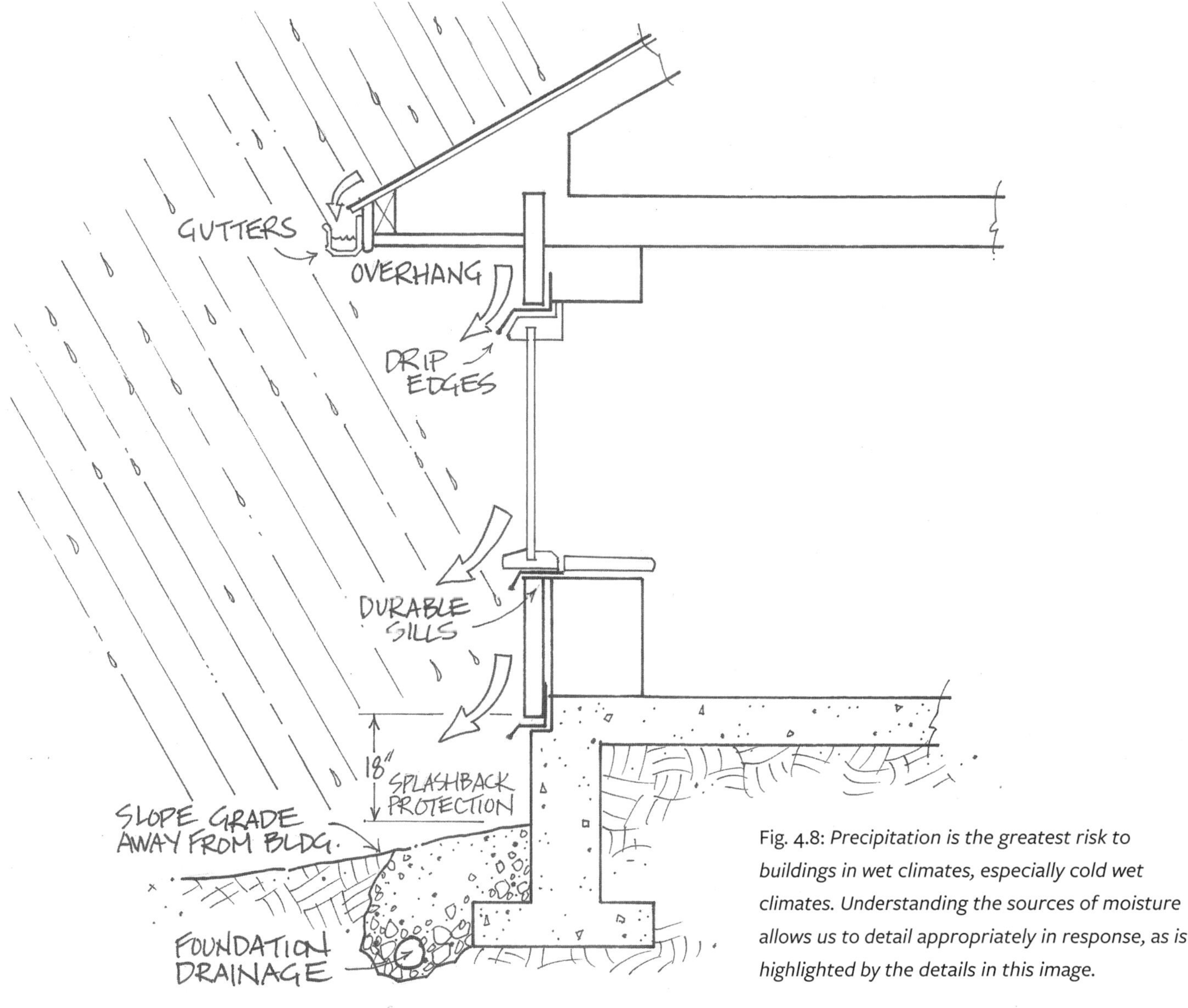

Fig. 4.8: *Precipitation is the greatest risk to buildings in wet climates, especially cold wet climates. Understanding the sources of moisture allows us to detail appropriately in response, as is highlighted by the details in this image.*

Splashback

Splashback is the wetting inflicted on the above-grade sections of foundations and bottoms of walls, where either driving rain or water falling from drip edges of roofs, doors and windows, and other architectural features hits the ground and "splashes back" against the building. Many factors influence the damage potential, including roof size/profile/material/overhang, gutters/diverters (or lack thereof), ground cover material and slope, and foundation/wall protection/height above grade.

Window/Door Leaks

The sheer magnitude of leaking doors and windows merits its own designation in our catalogue of rain and snow issues. These vital parts of our walls leak in many ways: through failed or missing flashing, through gaps and cracks between the units and the wall, into gaps below the sill through surface tension, through leaks and gaps in the units themselves due to poor installation, aging and poor-quality units, or faulty use.

Roof Leaks

The most common roof leaks occur as a result of poor flashing and protection detailing at transitions, such as valleys, dormers, and penetrations — the risk is proportional to the complexity of the roof in this regard. Another issue is ice-damming, in which frozen water at the eaves of a house stops liquid water from running off the roof, which then backs up under shingles and flashing and leaks into the house. This is discussed further in Chapter 6.

Surface and Groundwater

The ground can be an infinite source of moisture for our buildings, entering as *surface water* or as *groundwater.*

Direct Entry and Flooding

An easy way to identify surface water issues is to look for signs of water entry near the top of a basement, or around the perimeter of a slab. Staining on walls, silt trails, and cracks or blistering of finish surfaces are all indications of surface water. Surface water is generally attributed to specific wetting events, such as a sudden spring melt or a rainstorm. In fact, rainfall off a roof edge is a primary source of surface water infiltration. Groundwater will present primarily as dampness or puddling along basement slabs or lower walls, especially in areas of weakness or penetration, such as cracks, drains, or sumps, or in the center of a slab on grade. Groundwater is less tied to a specific event, and is more likely to show up as a seasonal pattern.

Water Movement into Walls, Framing

This is an extension of the issues highlighted above, but rather than remaining a basement condition, leaks into foundation walls or slab edges can wick upward into more sensitive parts of the assembly. The mechanism for moisture transfer is capillary action, as water is absorbed into the framing from saturated concrete. Whether that saturation is occurring close to grade from surface water, or down at the footings from groundwater, the capillarity of concrete could theoretically move water thousands of feet against gravity, by nature of its porosity.

Freeze-Thaw

Water expands when it freezes. The force of this expansion is incredibly powerful; it can even push buildings massively out of plumb (in the case of earth freezing and "heaving" below a foundation). On a smaller scale, water-saturated materials can be destroyed by expansion during freezing.

Vapor Entry and Movement

A large amount of humidity can make its way into a building from surface or groundwater sources either by saturating a foundation and then "drying into" the building, or through direct vapor transfer from saturated soil into the building by diffusion through voids in the foundation. Certainly the presence of puddling water in the basement can lead to acute damage, and even systemic mold and associated health issues. High humidity can be a major issue as well, in that this humidity can easily migrate through the enclosure and deposit in other parts of the building. It is not uncommon to track the source of condensation-related moisture damage in an attic back to high humidity levels in the basement, carried up through the building by way of stack effect and convective leaks into the attic. We discuss this in more detail later in this chapter.

Built-in Moisture

During the construction process, significant amounts of moisture can be built into enclosures. In many cases, this moisture will have the opportunity to dry out before the building is tightened up and finished. However, depending on the construction type, climate, season, sequencing, and chosen mechanical systems, serious moisture issues can develop within the first year of occupancy if this moisture isn't properly handled. Some of the ways moisture is built into our buildings include:

Curing concrete/masonry

A curing concrete slab or basement wall releases a lot of moisture into a building, and while it may cure well enough to build on within a month, it can take many months longer before new concrete is moisture stable.

Green timbers and wood framing/sheathing

Building with green lumber or timber (in this case, "green" refers to the moisture content, not the ecological footprint, color, or expense), or even kiln-dried lumber that has gotten wet, can be expected to be a source of built-in moisture. If you have an exposed interior frame, this moisture will dry into the building. If you have a wood frame within an assembly, it is wise to allow this frame sufficient time to dry — verified with a moisture-content meter to a common standard of 14% or lower — prior to fully enclosing, especially with materials that will limit drying potential.

Wet-install materials

Some materials require entrained moisture for their installation. Drywall joint compound, veneer plaster, and damp-spray cellulose insulation are good examples. In the world of natural building, we find a number of materials with significantly greater amounts of moisture, such as thick interior plasters and earthen floors. These applications may take a week or longer before they can be finished, and appliances such as fans and dehumidifiers may be required to facilitate timely drying. Bonded cellulose insulations — such as straw-clay, wood chip-clay, and hempcrete — may take even longer. If you are choosing a construction type that presents a significant moisture load, it is important to consider sequencing, factoring drying time into the construction schedule in such a way that the process isn't unexpectedly slowed down.

Managing relative humidity after close-in is important for any building in a cold climate. As construction continues after close-in, it is not uncommon for workers to use temporary heating systems in the form of unvented propane heaters; these generate huge amounts of vapor inside the building. If not dealt with, this situation can lead to mold and even severe rot in recently finished homes. Keep a hygrometer in the building, look for sweating on the windows or cold pipes, and be prepared to dry out a building if the RH gets too high (using dehumidifiers, exhaust ventilation, and circulation). It takes a lot of energy to deal with all this moisture — be prepared for elevated energy consumption (for heating, cooling, and ventilation) in the first year of occupancy while the building reaches moisture stasis.

Condensation and Humidity

Vapor itself does not have nearly the same potential to cause damage as liquid water; this makes sense, considering its concentrations are so much lower than in liquid phase. RH levels can be raised high enough through vapor diffusion alone to induce mold growth, but liquid water is needed for major decay to occur.

The greatest danger of water vapor is its ease of transport into sensitive areas of the building, where it can then undergo phase change into a liquid and achieve greater concentration. It is one thing to keep liquid from leaking into a building; trying to keep your buildings truly airtight, however, is a whole different story. In Part II, we discussed how warm air can hold large amounts of water vapor; that convective air cycles are driven along pressure gradients, making it easy for this warm humid air to be driven into our wall and roof assemblies; and why we might find liquid water in what we erroneously may have thought was a waterproof cavity. We know that this convective air leakage into our assemblies is the biggest vector for in-assembly humidity; to that, we can add diffusive vapor drive (not through air currents, but individual molecules of water vapor moving through permeable membranes) that can allow even more humidity to enter the assembly. The greater the humidity load in a building, the greater the risk to our assemblies. It is with this in mind that we should be considering the importance of addressing humidity issues.

The practical upshot of this humidity issue lies in the potential for condensation to form and damage sensitive parts of our buildings. In cold climates, we most commonly see this in the form of rotting roof decking and wall sheathing. In hot-humid climates, this is likely to occur behind wall and ceiling interior finish materials.

Condensation issues also arise where there are cold surfaces. Poor-quality windows may have interior condensation in the winter in cold and mixed climates, and exterior condensation in the summer in hot climates. Uninsulated below-grade surfaces are another area of condensation risk, particularly in the summer as a result of hot humid air hitting cold basement walls, slabs, and cold water pipes, and condensing. Opening your basement windows in the summer to "dry it out" is likely to have the exact opposite effect, depending on the relative humidities and temperatures of the outdoor air and basement.

Building durability and indoor air quality issues can both result from vapor migration/condensation issues, even when bulk liquid wetting is not an issue. Again, it is imperative that we understand the vectors through which this humidity enters our assemblies: humidity carried through air leakages is by far the largest source of vapor entry, and it is not appropriate to be worrying about vapor control without first being sure we've managed our air control. Vapor control is particularly important when employing the drying potential of the assembly.

Understanding the Layers

Let us now identify how the different control layers manage these different sources of moisture.

Water Control Layers

All the parts of the enclosure responsible for managing rain, snow, surface, and groundwater can be referred to as the *water control layer.* This layer may be located in different areas and composed of different materials, but is the most straightforward layer to understand conceptually. Waterproof membranes are applied toward the exterior of our roof, wall, and foundation assemblies to shed and repel water. Components such as flashing, gaskets, and sealants are used to handle transitions between different assemblies and breaks within the layer.

In roofs, this layer is generally a sheet membrane of asphalt-impregnated felt or plastic (or perhaps a layer of rigid insulation) located underneath the roofing, referred to as the roof underlayment. In walls, this layer is of similar material and is often referred to as the weather-resistant barrier (WRB), or *housewrap.* Liquid-applied membranes are also available, although these are less common in residential construction.

Note that for both roofs and walls, we are not considering the siding and roofing to be part of

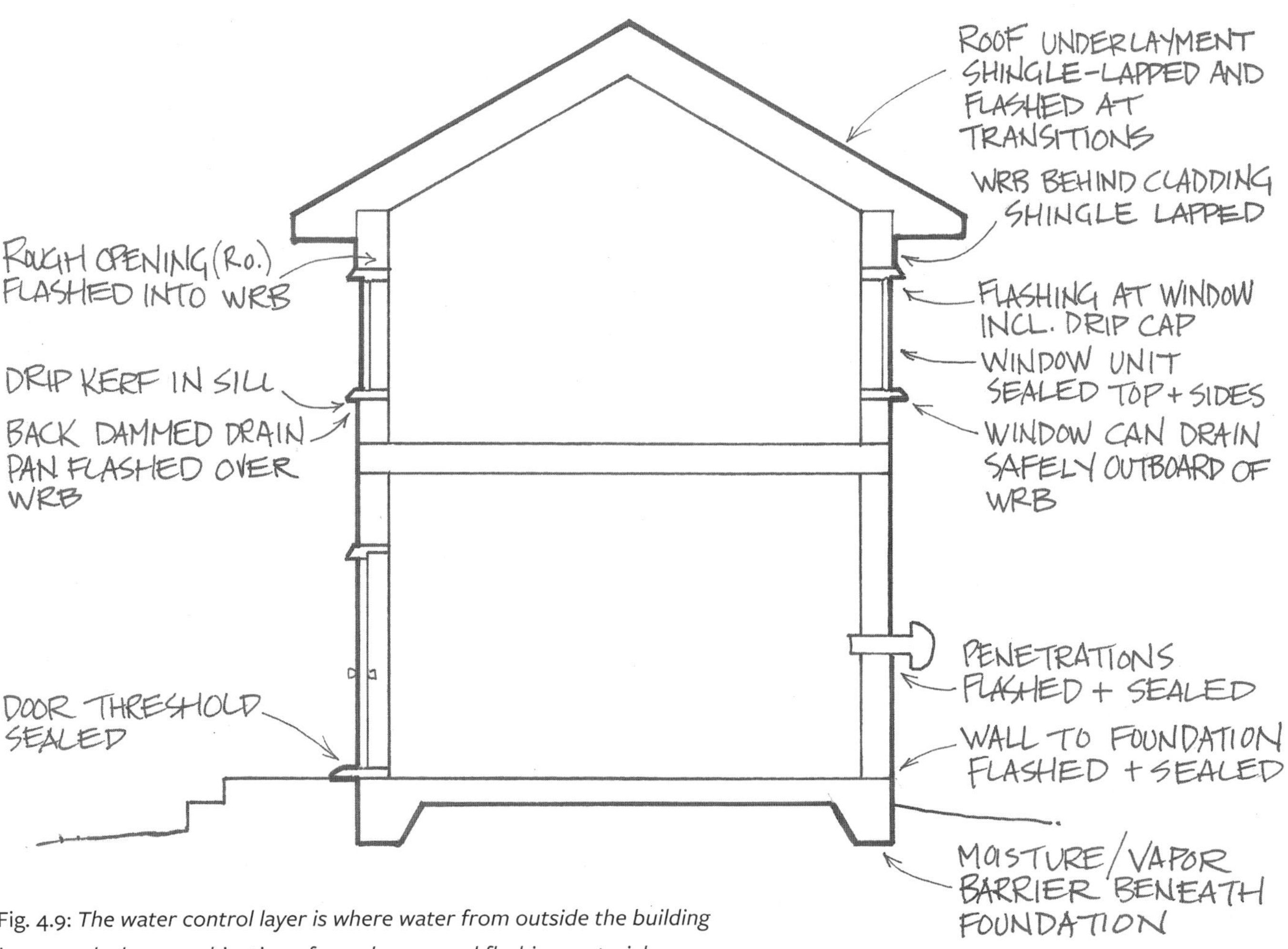

Fig. 4.9: *The water control layer is where water from outside the building is stopped—by a combination of membranes and flashing material.*

the water control layer, though these materials do provide a great deal of water control. Roofs, taking a tremendous amount of water impact, are very well-served by the redundancy of a backup layer; and, though siding may shed the majority of precipitation, we expect siding (and our windows) to leak and need to detail our assemblies with a water control layer, or even a drainage layer, behind the siding.

In foundations, things are a bit more complicated. The water control layer can take a number of forms, either as a liquid-applied barrier, a peel-and-stick membrane, or a sheet of plastic. Rather than rain, the water source here is ground or surface water. Much of this should be handled by drainage prior to the water reaching the foundation, but if the foundation is in contact with wet soil, the water control layer is critical to stop liquid diffusion and bulk flow from moving water into the walls or slab.

Air Control Layers

Our primary control layer for addressing condensation is our air barrier. This material may also serve as a vapor control layer if specified to do so, but not necessarily — many air barriers, such as plaster, wallboard, and airtight WRBs may be airtight but vapor-permeable. Again, "vapor" and "air" are not the same thing. Water molecules exist within a volume of air, which can contain very high concentrations of water vapor.

Just as our water control layers use different materials (membranes) connected by components (flashing, sealants) to create a

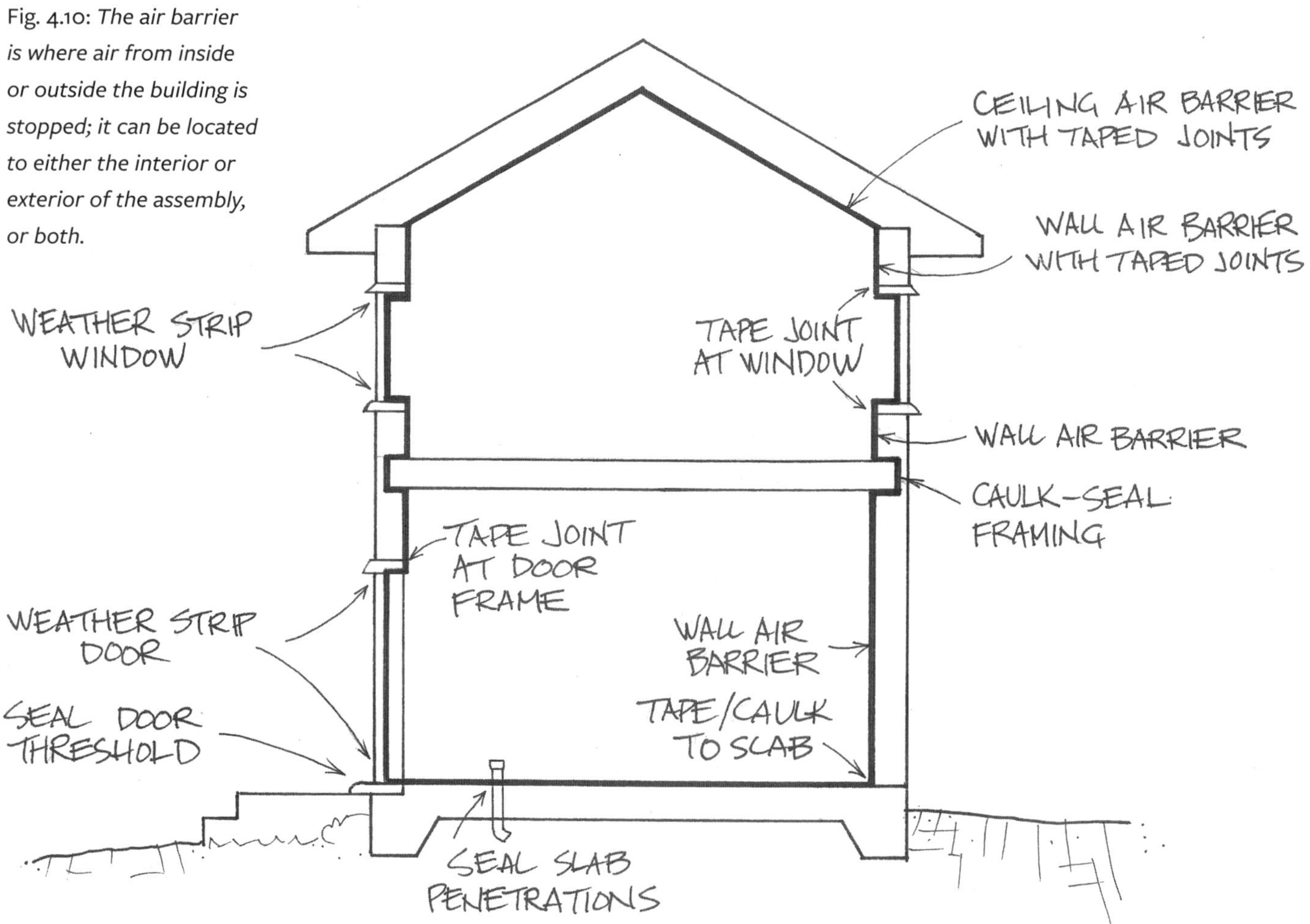

Fig. 4.10: *The air barrier is where air from inside or outside the building is stopped; it can be located to either the interior or exterior of the assembly, or both.*

fully functional system, so too must we think about how our air control layers are comprised. We generally use air barrier "materials" (such as plaster or a sheet membrane) to form the majority of this layer. These materials must be joined by "components" (such as caulk, tape, or gaskets) at all seams and at transitions to penetrations and changes in profile, such as windows, outlets, switches, and architectural features. These materials and components together create 2D "assemblies" (such as a flat wall or ceiling field), which in turn must all be connected at critical transitions to form a 3D "system" — the air control layer of your building.[5] Thinking of it in these terms helps to preserve a systems-based approach to the design, and helps us to focus on the transitions within the assembly or between different assemblies in the system, as it is in these areas where failures are by far the most likely to occur. In design, these areas are identified by taking a cross-section of a building and tracing your finger along the air control layer. Every time it hits a transition, that is an area to blow up into a detail. This may seem fussy, but it is necessary because of the pressure dynamic in buildings — the amount of air leakage into an assembly is not simply a factor of the size of the leak. Think of a garden hose: if you cover 75% of the end of the hose with your thumb, it will not reduce the amount of water coming out by 75% — most of the water still comes out, but at a much higher pressure. Accordingly, we need to chase air leaks down to a much higher degree of precision than one might initially assume, to avoid concentrating large amounts of air (and with it, vapor) in particular regions of our assemblies.

Unwanted vapor-rich air can enter an assembly from either the interior or the exterior, depending on the climate and the season. Our assemblies, particularly when they are well-insulated, tend to be quite thick. And, as we learned in Part II, convection occurs not just from air moving from inside to outside and vice versa (*exfiltration* and *infiltration*), but also air moving from the inside and back to the inside (called "re-entry loops"), from the outside back to the outside (called "wind-washing"), or even in loops within the assemblies (called "interstitial cycling").[6] We generally think of our air barriers only as stopping infiltration and exfiltration, and many assemblies are detailed this way. Accordingly, an air barrier is frequently put to only one side or the other of the assembly. If the insulation within the assembly is fully filling the cavity (if there is any cavity insulation) and is of sufficient density, this may be acceptable. If your insulation material is more porous, having air barriers on both sides of the assembly is important to ensure that the insulation will perform as intended, and to minimize the potential of condensation-related durability issues. Most building codes recognize this, and require fill and batt insulations in framed assemblies to be enclosed "on all six sides" to minimize air movement within the cavity.

The question of where to place the primary air barrier (the layer that connects to other assemblies to create a system) is of some debate. Some — myself included — prefer to locate the primary air barrier on the side where there is greatest humidity, to arrest airborne humidity entry at its source (this would be to the interior in cold climates, to the exterior in hot-humid climates). Others always prefer the outside, as it is easiest and cheapest to install (fewer penetrations and obstacles compared to the inside) and prevents wind-washing, while some always prefer the inside (easier to access and repair — potentially — and to connect to ceiling and foundation layers). Ultimately, you may well need two — for example, a primary air barrier to the interior, and a secondary air barrier, or "wind barrier," to the exterior, which may not tie into the air barrier materials of the adjoining assemblies, but will prevent wind-washing in the cavity.

If you are designing a 100-year (or longer) assembly, it is imperative that the air barrier be designed in such a way as to ensure its protection and maintenance over the lifespan of the assembly. These are often competing values — some designs favor burying the air barrier in the assembly behind the "services" of electricity, telecom, ventilation, etc. (creating a *service cavity* inboard of the air barrier), which minimizes future disturbance but makes access difficult. Other approaches favor having the surface material of a wall or ceiling, such as drywall or plaster, act as the air barrier; this can ensure long-term visual inspection and maintenance, even as buildings change occupants and experience remodels.

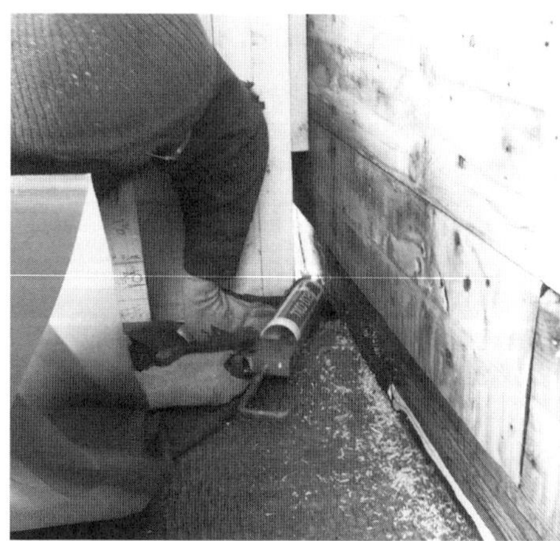

Fig. 4.11: *A high degree of thoroughness is required to achieve an airtight assembly, but the goal can be quite attainable if considered early in the design process.* CREDIT: TOP AND RIGHT: ACE MCARLETON — NEW FRAMEWORKS NATURAL DESIGN/BUILD.

However, by virtue of their exposure, they are at greater risk for intentional or unintentional damage that would compromise their performance. In either case, communication with the owners — in the form of written documentation such as an owner's manual that can transfer along with ownership — is a critical component of the design of a successful air barrier; the owner is the final "stakeholder" who must be on board to ensure long-term viability.

We have the know-how, materials, and tools to easily build robust, durable, relatively airtight assemblies with little to no additional cost. But there are a few more critical — and often overlooked — issues involved, including coordination with other stakeholders, design detailing and testing procedures that ensure the work is being done well. Many different trades may all potentially be interacting with this layer, and they may not be accustomed to thinking about airtightness. Not only must everyone on the site be on board, but there must be terrific communication between the designers and the builders to ensure not only that the details are well-defined and described (builders listening to designers) but that the details are practical, affordable, and feasible (designers listening to builders).

Regarding the air barrier "components": it is often at the transitions where failures occur. Always check for chemical compatibility and follow the manufacturer's specifications to the letter. Further, installation details matter greatly, as there are right and wrong ways to install caulk; many different gaskets serve many different purposes, and particular conditions are required for a "permanent" tape joint (dry, clean, warm, stable substrates). If the wrong product is installed, or the right product is installed incorrectly, it will be the weak link in the air barrier system.

Above: The design of the framing must be considered when building a high-performance enclosure to ensure thermal bridging is minimized. While the role of framing is to provide the structure, it also must interface with the layers that control the transfer of air, water, and heat.

Left: Wood fiberboard sheathing is used on the exterior of the framing to provide enclosure, structure, and a bit of added insulation. Fiberboard sheathing allows good exterior drying potential, does not quickly degrade in the presence of liquid water, provides R-4/inch of insulation, and has a low embodied carbon profile.

Top: Straw is used in combination with cellulose (dense-packed in the framing cavities outboard of the straw) to create a super-insulated wall. An airtight highly insulated front door was specified to match the performance of the adjacent walls.

Above: Round timbers complement the curvature of an interior straw bale corner, but can present challenges for easy air-sealing where the framing interrupts the plaster, and a joint must then be managed.

Lower Right: The availability of durable air-barrier tapes has made achieving airtight construction much easier. Fleece tapes are available for plaster-to-wood transitions.

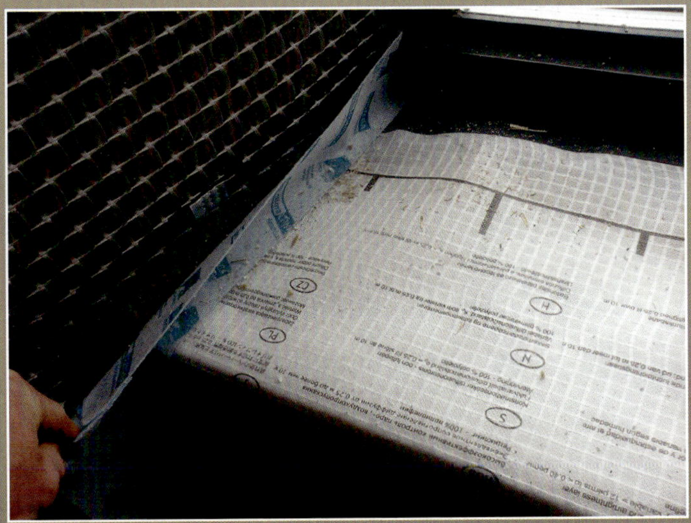

Top Left: *When it comes to building airtight construction, detailing the transitions is the most critical part of the process. Here, a piece of fleece air-barrier tape is installed to secure a connection between an air-barrier membrane over a window interior sill and the adjacent plaster wall finish.*

Top Right: *Thick walls create great opportunities for window seats. High-mass materials such stone, concrete, or tile, can be used as interior sills for southern windows to store passive solar heat in northern climates.*

Bottom: *Beauty is a critical part of sustainability; regardless of the R-value and water management detailing, an ugly building won't satisfy its users, and won't be well cared for over time. There need not be a conflict between form and function with a well-designed home!*

Above Left: *Flashing is designed to force water to drain safely away from vulnerable transitions and protect sensitive materials deeper within. Bottom-of-wall detailing is of particular importance, as the exposure to moisture in wet climates is high in an area with critical transitions from foundation to wall.*

Above Right: A vent cavity is created behind the shingles of this curved roof, both allowing the shingles to dry out (preserving their longevity) and encouraging outward drying potential of the roof assembly.

Right: At every transition and every edge, water has a potential to gain entry into the assembly. The lifespan of the building may well be determined by the successful execution of these transitions, from design to construction detailing.

Top: *Building high-performance enclosures and mechanical systems alone isn't enough to make a good house. Consideration of the context – the climate, the architectural vernacular, and the regional resources – is critical to ensure the building is responding to and supportive of the environment in which it is built.*

Left: The mass of native stone – reclaimed from the foundation of a barn on the property – is used to help capture the heat emitted from the wood cookstove to temper the heating curve and continue to slowly release heat into the night. Wood space heating is backed up by solar and wood-powered hydronic heating to provide the minimal heat load.

Top Left: Thick walls are a hallmark of many high-performance homes, offering design opportunities for interior sills and window seats. All non-toxic finishes help preserve the air quality of the house – critical with airtight construction.

Top Right: Whole systems design informs this high performance home, featuring a foam-free airtight super-insulated enclosure, passive solar design, and fossil-fuel-free mechanical systems, including PV and solar hot water, a pellet boiler, and ductless HRVs.

Lower Right: Two different zones of radiant heat tubing run beneath the natural wood and tile sections of the floor. The zone below the wood runs at a lower temperature to avoid splitting, with hotter water circulating beneath the tile to help pick up the rest of the heating load.

Left: Vertical board and batten siding makes good use of regional wood to provide a durable rainscreen siding; working with local wood in a vernacular style was a priority for the owner.

Above: On the north facade of this owner-built building, fewer windows are present to reduce thermal loss. The simple design was cost-effective and achievable for an owner-builder to execute without sacrificing the performance of the enclosure.

Left: With the StrawCell straw/cellulose hybrid wall system, the durability of a well-drained rainscreen on the exterior combines with the performance of a high-mass airtight interior plaster over the straw bales to create an exceptional super-insulated wall assembly.

Top: This 19th-century Vermont farmhouse was brought into the 21st century by insulating and air-sealing the basement, walls, and attic, replacing the windows and doors with triple-pane fiberglass-frame units, and installing a full PV system, ducted HRV, and a geothermal heating and cooling system.

Above Left: By replacing the existing vinyl siding, an opportunity was created to wrap the walls with additional insulation and an airtight water control layer, enhancing both thermal performance and durability.

Bottom Right: This custom-built and designed high-performance door features local wood, mineral board insulation, and thermal-bridge-free R-9 construction.

Testing for Airtightness

The blower-door test is the industry's standard tool for assessing the quality of the air barrier, as well as for quantifying the overall leakage in a building. During the test, the building is shut tight, all combustion equipment is turned off, and a temporary shrouded frame is set up in a doorway with a large fan hung in the frame. Hoses are connected from the outdoors and from the fan to a pressure gauge, called a *manometer.* The fan is turned on, blowing air out of (or less frequently, into) the building until a pressure difference of 50 pascals (a unit of pressure that's equal to about 1 pound per square foot) between the inside and the outside is reached. The gauge then reads the amount of air blowing out of the building. Think of a balloon: if the balloon doesn't have any holes, you only have to blow it up a little bit to keep it filled. If there are lots of pinholes in the balloon, you have to blow in a lot more air to keep it fully pressurized!

The manometer is able to calculate the number of cubic feet of air per minute (CFM) required to be blown out of the building to depressurize the building to the 50 pascal pressure difference; the bigger the number, the greater the leakage. The results of a blower door test are therefore expressed as "CFM50," or the number of CFM at a standard pressure of 50 pascals. This test simulates a 20 MPH wind on all exterior surfaces of the building. With the blower door on, we are able to inspect the building and scan for actual air leaks. The air changes per hour at 50 pascals (ACH50) represents the number of times the total volume of air in the building leaks through the enclosure under the testing conditions; it provides a baseline for comparison between houses or for code compliance. For example, the International Energy Conservation Code (IECC) 2012 prescribes a limit of 3 ACH50 in climate zones 3 and above. A very tight home might be less than 1 ACH50, while a drafty old farm house might be 8 ACH50, or higher. Another metric, CFM50/exterior square footage, looks at how leaky the building is — not compared to its volume, but compared to its surface area. Since it is through the building's surface area that air leaks into or out of the building, this is perhaps a more relevant metric, although it is less commonly used in the residential world. It is also worth noting that the volumetric ACH50 value does not compare well across buildings of different sizes or shapes, as large and simple buildings tend to have a lower surface area-to-volume ratio, and therefore test "better" than smaller or complex form buildings. However, ACH50, converted to a "natural" air changes per hour value, allows energy modeling to take into account heat loss or gain through expected air leakage.

We use the blower door frequently during the building process. During new construction, we use the tool at least once, if not multiple times, to make sure we are on target with our air-sealing goals; this allows us to correct the installation of all the various materials and components prior to finishing. While the blower door is running, we can use smoke, infrared (IR) thermography, or the backs of our hands to identify leaks in our layers and then verify their sealing. Sequencing the construction process to allow for this quality control is an important consideration during the design and planning phase. We also do a final test prior to move-in to verify our as-built air leakage level in the building. In retrofits and renovations, we do a baseline test prior to conducting any work, so that we can quantify the level of improvement over the course of the project. Testing is then conducted mid-stream and at the end of completion as per new construction, often in concert with IR thermography according to best common practices.

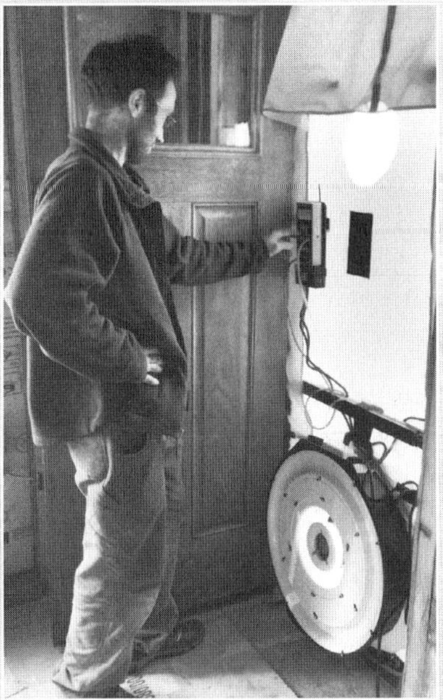

Fig. 4.12: *Diagnostic tools such as a blower door test are critical for builders to ensure that the air control layer is installed to specification, and to quantify the overall air leakage of the building.* CREDIT: JACOB DEVA RACUSIN — NEW FRAMEWORKS NATURAL DESIGN/BUILD.

Vapor Control Layers

Now that we've worked through our air control layer, we can start addressing the vapor control layer in our assembly. Vapor control can be a pretty complicated topic, especially compared to air control. With air, we can say there is no reason to encourage airflow into our assemblies, and that our air control layer either functions as an air barrier, or it doesn't — there is no gray area in between. Not so with vapor control. Some assemblies rely on allowing vapor to flow all the way through (called *flow-through* or *vapor-open* assemblies); some allow vapor to move only to the interior or exterior; and still others don't allow vapor flow at all. In these various scenarios, the location of the vapor control layer becomes critical, and it can be the difference between success and failure. Further complicating matters, there is a wide range in vapor permeability among different materials, and some materials are *vapor variable,* meaning their vapor permeability changes based on the relative humidity or material moisture content. (Wood is a good example of this: when dry, wood is semi-impermeable, but it becomes more permeable as it gets wet, promoting its ability

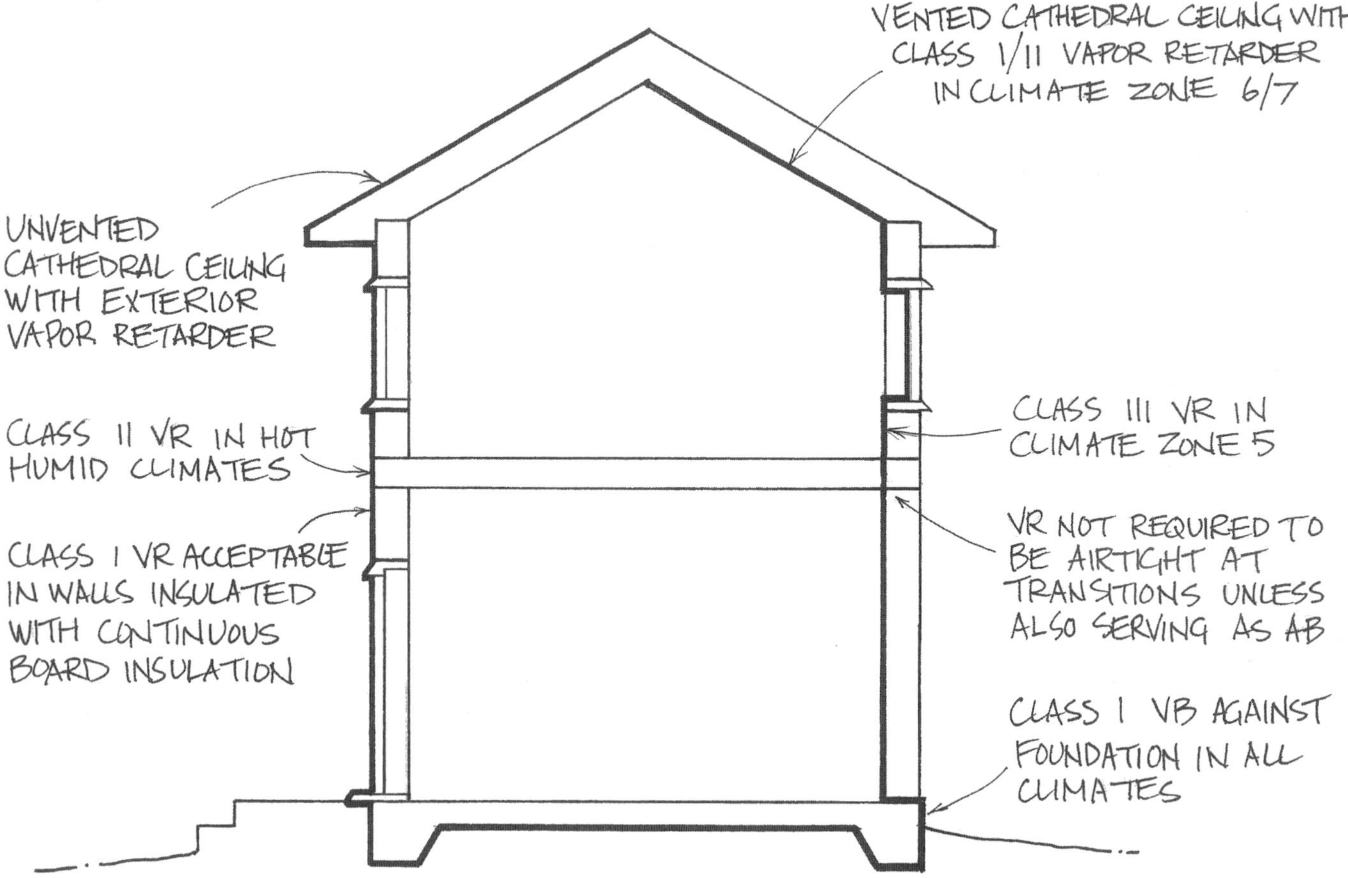

Fig. 4.13: *The vapor control layer is where vapor from inside or outside the building is slowed or stopped by a vapor control layer, or retarder, which can be located to the interior, the middle, or the exterior of the assembly, depending on climate.*

to dry out.) And while air control strategies are easily replicable across different climates, vapor control strategies must be tuned to the climate conditions of the building. Vapor control is one of the most climate-dependent components of our building enclosure design, and it requires a regionally appropriate approach. A thorough understanding of how to design for all of these variables in all contexts is well beyond the scope of this book, but we will attempt to define and contextualize the variables.[7]

To organize and quantify the scale of permeability of different materials, building codes recognize three different categories of vapor retarders, based on their permeability. Table 4.1 presents the different classes of materials along with their descriptions and their permeability ratings (expressed as US perm, or 1 grain water vapor per hour per square foot per inch mercury — a pressure metric) and example materials for each class, as given in the 2009 International Residential Code (IRC).[8] Note that permeability is assessed using either *dry-cup* or *wet-cup* methods, referring to the relative humidities on either side of the material during the test ("dry-cup" tests for permeability when the RH is low, and "wet-cup" tests for permeability when the RH is high). As discussed above, the permeability will vary between dry-cup and wet-cup conditions for certain materials, and the type of test must be confirmed when specifying a given material, or comparing two materials.

The role of the vapor control layer is to limit vapor diffusion through the materials — *limit* is the operative word here, as we don't necessarily want to stop vapor diffusion. Why not? Because vapor moves both into and out of our walls; if we put a vapor barrier in our wall or roof, we are stopping vapor from diffusing in, but we are also stopping it from diffusing out, eliminating the *drying potential* in that direction. This can be a

Table 4.1

VR Class	Description	Permeability	Materials
Class I	Vapor-impermeable (vapor barrier)	≤ 0.1 perm	Sheet polyethylene, sheet metal, aluminum foil, foil-faced insulation, rubber membranes, glass
Class II	Vapor-semi-impermeable	> 0.1 to 1 perm	Kraft-faced fiberglass batts, oil-based paint, vinyl wall coverings, XPS foam ≥ 1″ thick, cement stucco
Class III	Vapor-semi-permeable	1 to 10 perm	Plywood, OSB, EPS foam, fiber-faced polyiso foam, most latex paint
Vapor-open	Vapor-permeable	> 10 perm	Gypsum wallboard, earthen and lime plaster, asphalt-impregnated felt, most housewraps

The Danger of Radon

Because we are discussing vapor, this is an appropriate place to mention another problematic gas — radon. Radon is a carcinogenic gas found in soil and groundwater, most commonly in regions featuring granite as part of their geologic parentage. Radon can be a significant health hazard, particularly in buildings with basements or unvented crawlspaces. A full discussion of radon is beyond the scope of this book, but if you are building in a region known to have radon in the ground (though radon levels vary widely from site to site), you must seek to control it. Many sub-slab vapor barriers are also rated as radon barriers; other strategies involve exhausting the area below a slab through a fan connected to a drainage pipe. Do not fail to test for radon in an existing home; in new homes you should install a radon mitigation system. Studies have shown that, even in high radon areas, a passive (stack effect) radon vent is sufficient to mitigate radon concentrations in most homes. The International Residential Code requires a passive radon mitigation system in most new construction.[9] We must note that radon isn't the only soil gas that might be of concern in our buildings. It is one of the most dangerous, but methane is also an issue in some areas; 'sour gas' from fracking activities is making this more and more common. Other gasses may be problematic as well, depending on the region.

mistake. Preserving the ability of our assemblies to dry to one side or the other is a fundamental (but not universal) strategy for good moisture management.

A big distinction between an air barrier and a vapor barrier is that while small holes in an air barrier can channel through a lot of air in a small location, the effectiveness of a vapor barrier is proportional to its surface area. Covering 95% of a crawl space floor with a vapor barrier will result in a (roughly) 95% reduction in vapor diffusion.

One of the most fundamental design principles for building enclosures is to ensure they are complete and explicit. "In-between" spaces that are partially outdoors or partially conditioned, such as basements, attached garages or mudrooms, enclosed porches, and "cold" attics, often lead to confusion about where a given layer begins or ends (for example, when the insulation stops at the main house walls but the attached mudroom is kept above freezing). Each layer must be drawn clearly and completely, even if the layers split (for example, the thermal control layer stops at the main house, but the water control layer wraps the perimeter of the mudroom). This discipline of clear and consistent layers in design is carried through in construction detailing. It may be simple to identify the layers in a representative cross-section in the middle of a wall, but it is understanding which layer connects to what material, and how — where a wall meets a window, or a ceiling or a floor — that determines the quality of an enclosure. This guiding principle will inform the quality of the enclosure, from the thoroughness of the design process to the quality of the execution of the build — so don't skimp on the details!

Moisture Balance

When we incorporate all the elements of our moisture control strategy effectively, we achieve a *moisture balance* in our buildings. As opposed to simply trying to keep all the moisture out, coupled with a half-way nod toward drying, a more holistic approach recognizes that vapor is regularly entering, staying for a while, and then leaving our buildings (with much of it created by the occupants inside); this can even be the case with liquid water for short periods of time. Our job as designers and builders is to recognize the sources of this moisture and make sure that our assemblies can dry out faster than they are wetted and can safely store moisture long enough to dry before damage occurs. The most durable and low-risk walls and roofs keep these factors in continuing balance.

Let's look at a moderate-risk wall assembly to illustrate the concept of the moisture balance: an airtight double-stud wall insulated with dense-pack cellulose featuring board sheathing and a rainscreen on the exterior and latex-painted drywall on the interior (so, no vapor control layer) in a cold climate. A lot is being done to keep the moisture out: a rainscreen with good detailing keeps out the rain and snow; the airtight design keeps out much of the vapor; and a good dense-pack minimizes convective cycling in the cavity. Storage is achieved by the cellulose and by the wood sheathing, and drying can be achieved in either direction. However, in a cold winter, that sheathing gets very cold, and there is no check on the vapor drive from the building interior through the assembly. Will the sheathing be able to safely store the condensed vapor long enough before it can dry fully to the exterior? If the RH in the building is kept in a safe range (say, below 40%) and no other significant wetting is introduced into the system, yes. But what if the RH in the building rises to 60%? A vapor control layer would likely be required. What if you swap out the wood sheathing for OSB? The ability of the material to tolerate a higher moisture content is now inadequate, and the sheathing will likely be damaged; the drying capacity would also be reduced, as the wet-cup permeability for OSB remains relatively low. This condition would be made even worse by swapping out the cellulose for fiberglass batts, which would both reduce storage and allow for more convective activity, resulting in more wetting. Remove the rain-screen, and the drying potential is reduced. Cut a bunch of holes in the drywall without patching them to an airtight level, and spot vapor loading through air infiltration will create too much wetting. This is not to suggest that this is an unsafe wall assembly, even in a cold climate. By understanding the philosophy of achieving a moisture balance in our assemblies and our buildings, we can arm ourselves with the perspective to critically think our way through decision-making on the material and detail level — and that's where we find all the action.

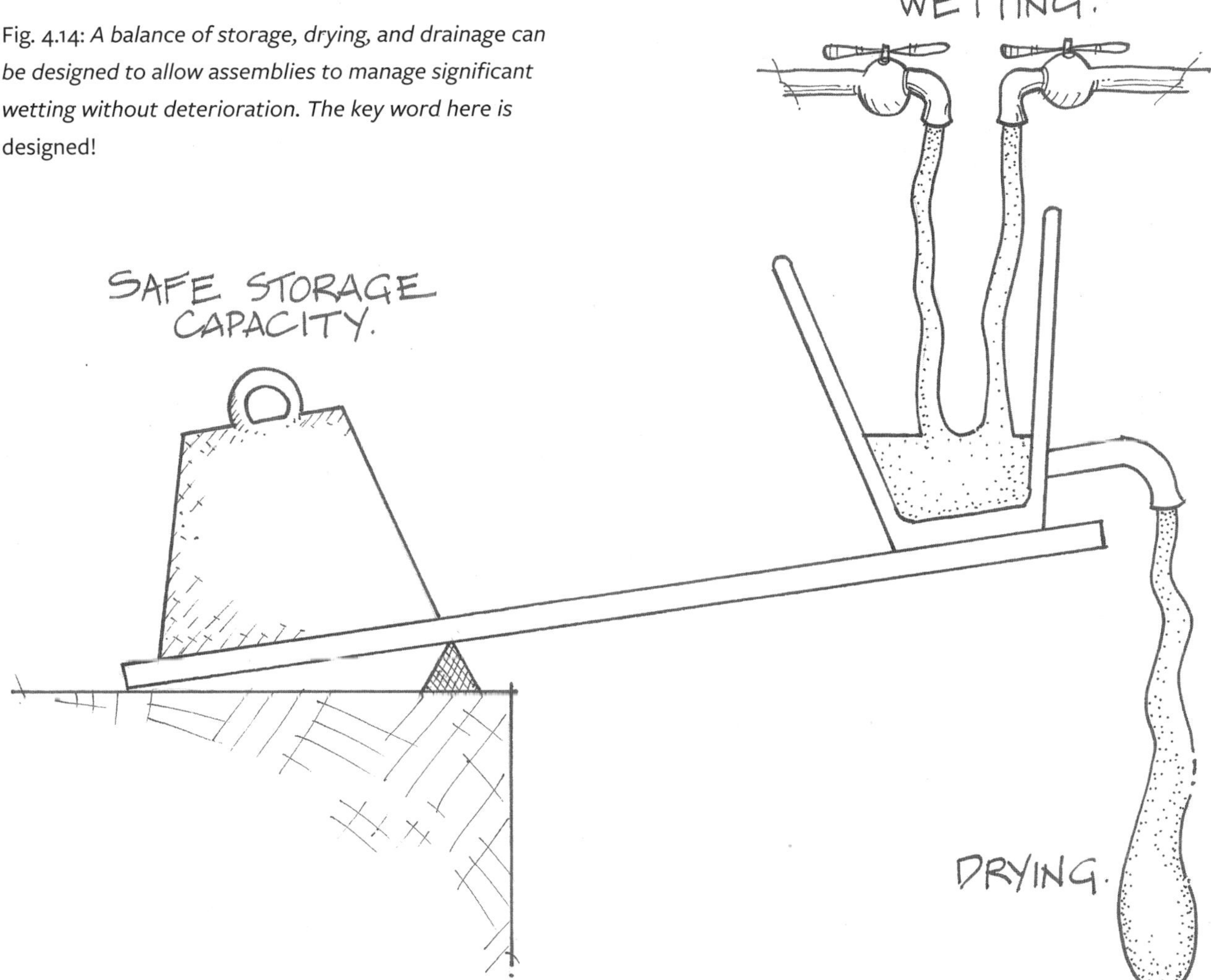

Fig. 4.14: *A balance of storage, drying, and drainage can be designed to allow assemblies to manage significant wetting without deterioration. The key word here is* designed!

Chapter 5

Thermal Control

Now that we have unpacked how the dynamics of moisture play out in our buildings, we can do the same with heat. To achieve the objective of keeping heat in or out as desired, we turn to our thermal control and air control layers to do the job.

The Thermal Enclosure

The *thermal control layer* — our insulation — interfaces with the air control layer to create our *thermal enclosure.* There are four rules for a quality thermal enclosure:

1. There must be insulation (thermal control layer).
2. There must be an air barrier (air control layer).
3. These layers must be touching each other.
4. These layers must be continuous.

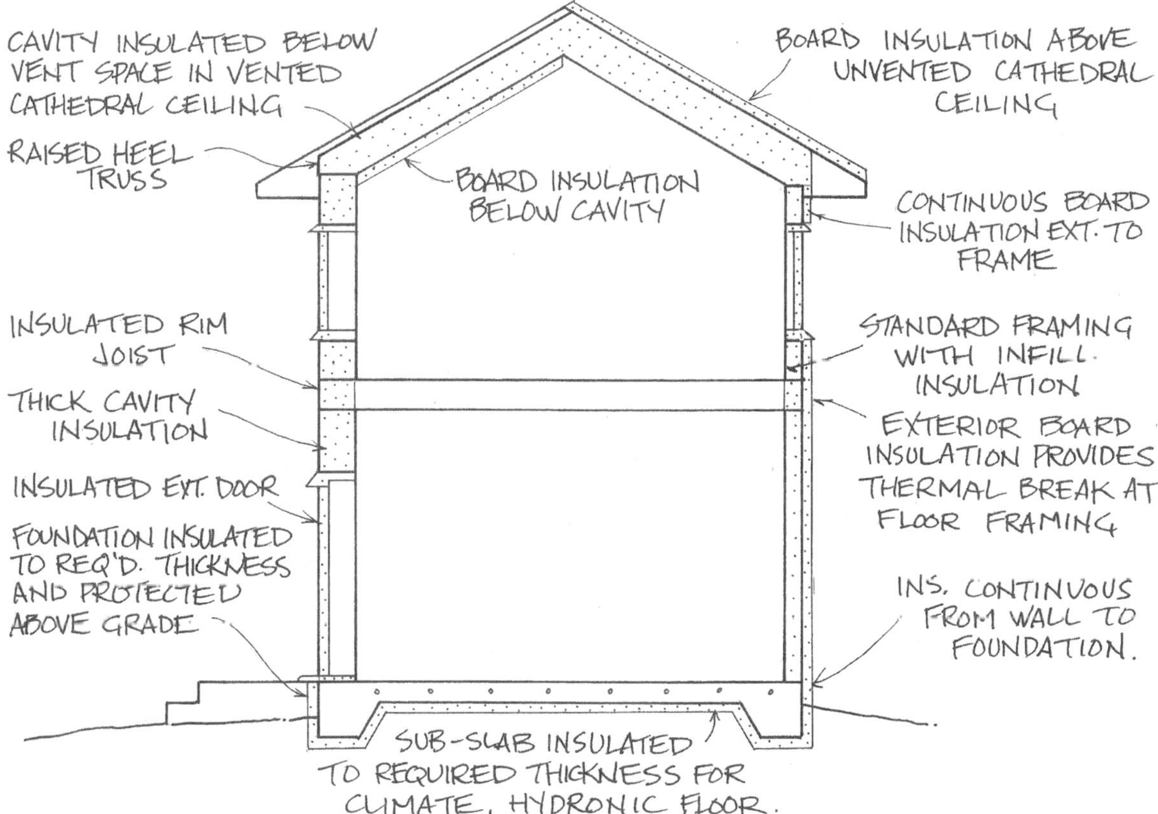

Fig. 5.1: *The thermal control layer is where heat energy from inside or outside the building is slowed (it is never truly stopped) by insulation, which can be located to the interior, middle, or exterior of the assembly.*

Sounds so simple, right? As with most things, the practice is more difficult; transition details can be problematic and complicated. And just as with the other layers, it is critical to work out these details in design, as it is often too late, and certainly much more expensive, to try to rectify problems in the field after the fact. Every additional roof valley or hip, wall corner, overhanging floor, bay window, dormer, skylight, and penetration increases design and construction cost, risk of failure, potential compromise in performance, material waste, and maintenance liability.

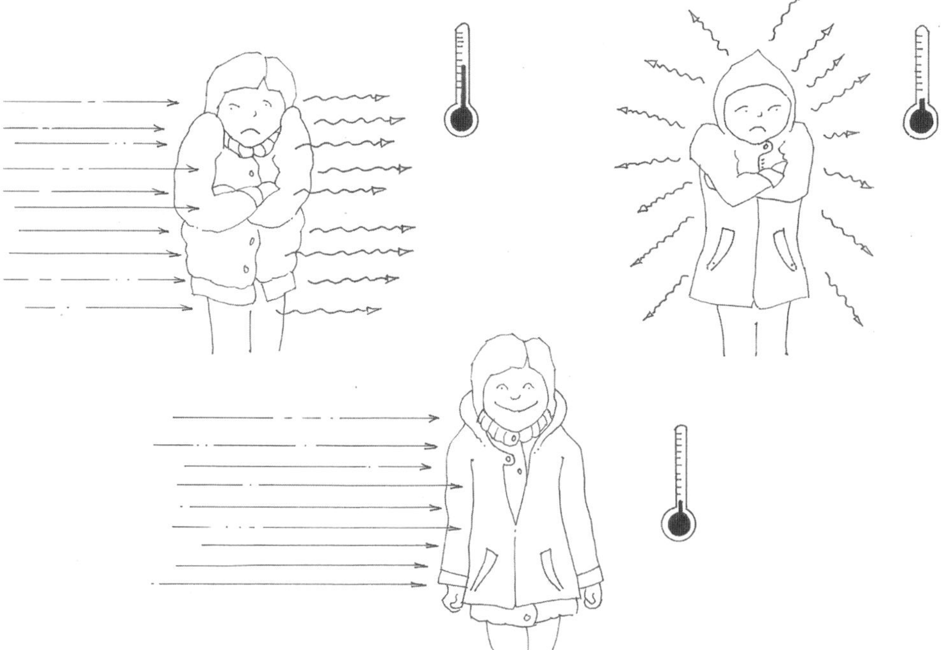

Fig. 5.2: *Think of the thermal enclosure like dressing for the winter. If you wear a sweater with no wind-breaker, the wind whips in and freezes you. If you wear a wind-breaker with no sweater, you lose a lot of heat through the thin shell. But wear both, and you'll stay toasty. And if it's a vapor-permeable wind-breaker made out of Gore-Tex®, you won't get sweaty—as you would wearing an impermeable rubber raincoat!*

About Radiant Barriers

One of the most confusing products on the market is the radiant barrier. These shiny-faced materials are frequently peddled as miracle products with astoundingly high R-values, with proven use in space by NASA. If it sounds too good to be true, it probably is. Radiant barriers are designed to retard the flow of radiant heat through an assembly. The shiny surface has a very low emissivity and high reflectivity, meaning it will reflect radiant energy on the hot side of the barrier, and reduce radiant energy emission on the cold side. In order for these materials to work, there must be an air space on one side or the other, because radiant heat moves through space. Generally, 1 inch (25 mm) is recommended; thinner spaces allow conduction to overwhelm radiation as the primary transfer route. If there is solid contact on both sides of the barrier, the whole unit simply becomes a conductor. One of the best applications for radiant barriers is on the underside of roof decking or roof framing in hot climates, where radiant energy from the sun hits the roofing, conducts through the materials, and attempts to re-radiate downward; but it is then stopped by the radiant barrier, reducing the amount of heat transfer to the unconditioned attic inside. But if someone tries to sell you an R-20 radiant foil barrier to put under your slab (where it is practically impossible to create an air gap adjacent to the radiant barrier), ask for independent laboratory test results — or just keep walking.

The role of the thermal control layer is to slow unwanted heat loss or gain through the building assemblies. We use insulation and, in limited cases, radiant barriers to achieve this; our windows and doors end up playing very important roles as well, since punching holes in our enclosures tends to have a negative effect on thermal performance. Figure 5.3 helps explain how insulation works.

There are many factors that go into a building's thermal performance, one of which is the physical quality of the insulation itself. Insulation materials are rated by R-value here in North America, which measures the resistance to heat flow and is given in units of

---

While the stated R-value of a material may be a helpful reference for comparison and even for modeling and predicting performance, it is not the only consideration in our evaluation of the thermal performance of a building.

Fig. 5.3

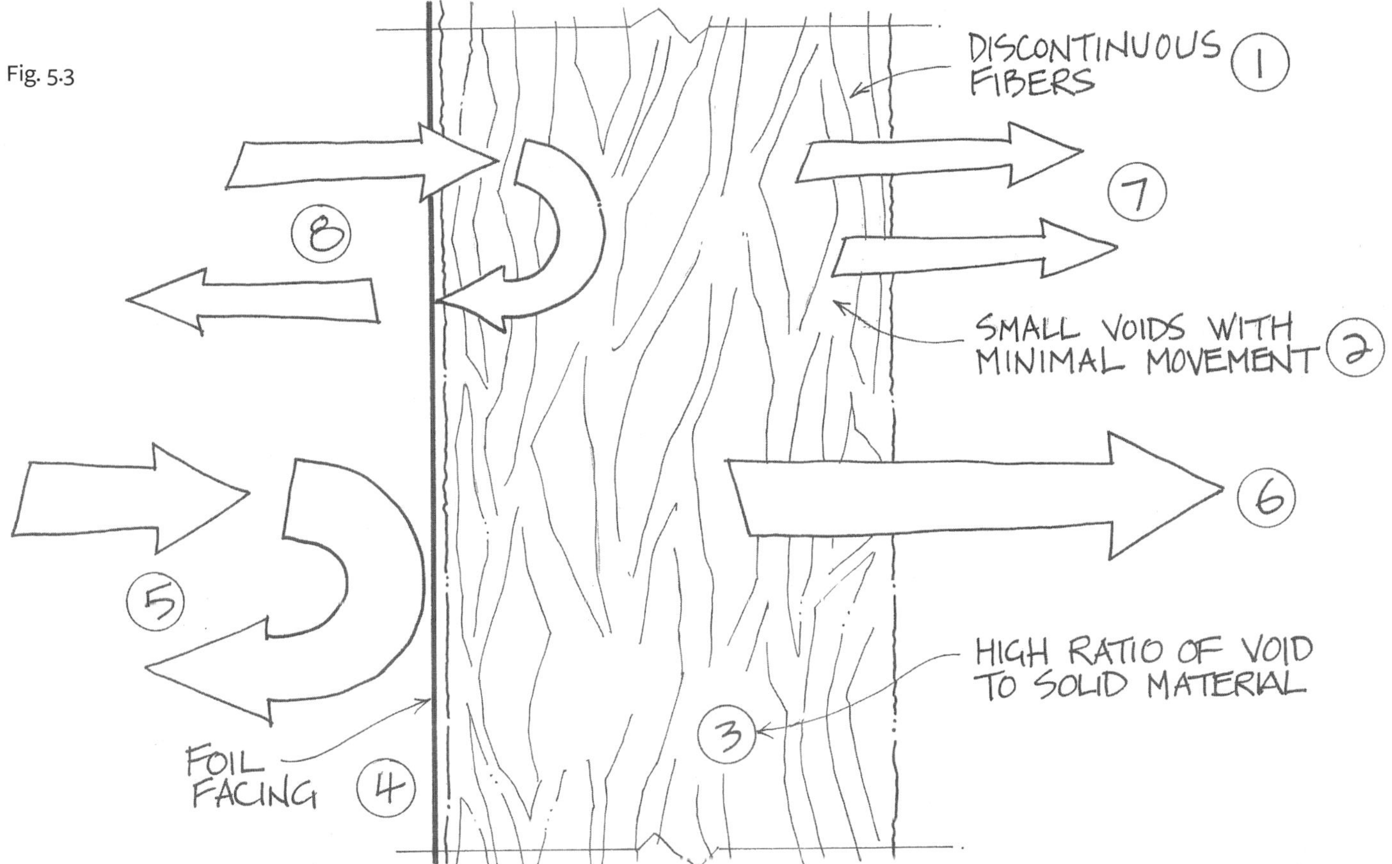

1. Disconnected fibers, foam, other insulation material.
2. Tiny air or gas spaces, disconnected from each other, minimal movement between spaces.
3. High ratio of void spaces to solid material.
4. Radiant barrier.
5. Radiant energy is reflected back by highly reflective radiant barrier.
6. Limited amount of energy is transmitted through the material and emitted through the assembly.
7. Conduction is greatly slowed; limited amount of energy conducts through the assembly.
8. Convection is minimized within the insulation, and between voids in the insulation when the cavity is fully filled.

ft²·°F·hr/Btu (m²·°C/kW). This is the inverse of the U-value, which is a measure of how easily heat flows through a material: R-20 = U-(1/20) = U-0.05. We generally evaluate a material on an R/inch basis (i.e., dense-pack cellulose is R-3.5/inch), and assemblies as a total R-value (i.e., an R-40 wall). R-value tests for insulation materials are conducted under controlled conditions as prescribed by various ASTM testing standards — namely, warm, dry, and still. In the real world, however, when we want our insulation to be doing its job, conditions are not warm and still — it is blazing hot, freezing cold, very windy, and/or wet from condensation or leakage. Further, the materials don't work in isolation; they work as part of multi-layered assemblies, with other less-insulating materials, such as framing, built into the mix. To that end, while the stated R-value of a material may be a helpful reference for comparison and even for modeling and predicting performance, it is not the only consideration in our evaluation of the thermal performance of a building. Rather, we must put these materials in context: how well do they function as part of an assembly — when it is cold or wet or hot? What other contributing factors in our assemblies affect this performance, such as framing, air spaces in the cavity, or wetting potential of our assembly? And how do different materials hold up over time? These variables matter, and you should pay attention to them. One of your primary goals for a high-performance enclosure should be minimizing or eliminating thermal bridging; another is maximizing real-world thermal performance by selecting the correct type of insulation to perform in the context of the assembly.

Fig. 5.4: *Insulation only works where it's placed. With no insulation breaking the framing from the exterior (or interior), the framing will act as a relative conductor, or "thermal bridge." There is little room to insulate in the condition pictured here; this is a problem created in design.*
CREDIT: JACOB DEVA RACUSIN — NEW FRAMEWORKS NATURAL DESIGN/BUILD.

Insulation Options

In surveying our many insulation options, we can evaluate them based on criteria such as durability, environmental impact, permeability, cost, effectiveness, and ease of installation. Given the importance of making sure that the insulation works in the context of the assembly to achieve its performance potential, it is important to consider the physical form: rigid, batt, loose fill, dense-pack, or spray-applied. While there are many types of insulation, we will attempt to catalogue some of the most common forms, with brief commentary regarding their thermal performance, durability, cost, ease of use, health impact, and environmental impact.[1]

Rigid Board and Block Insulation

Bio-based boards include fiberboards (made from wood fibers) and cork, (made from the bark of cork trees). Most are not structural, and none are suitable in wet locations (i.e., below grade). Both fiberboards and cork have fire ratings better than foams, but worse than mineral board.

Mineral board (also known as mineral wool/fiber board) is often used as a replacement for foam for those concerned about ecological impact, health or fire risk. Made from spun blast furnace slag, these boards are totally fire and moisture-proof, and, like most foam, continue to perform at their rated values in wet (but not saturated) conditions. They contain far fewer chemicals, but formaldehyde is used as a binder, which is a human health hazard, and the particles are irritants to skin and lungs.

Foam boards are ubiquitous. These insulations can serve as both thermal control, vapor control, air control, and water control layers, if detailed properly. Foams are amongst the best insulating materials, however the R-value of polyiso is known to degrade over time due to leakage of blowing agents (the chemicals that create the gas spaces within the foam), and they degrade further under cold conditions[2] — many authorities use R-4 or R-5 in cold service, and R-6 in warm service locations. Plastic foams have a high global warming potential and are the source of ecologically persistent toxins.

Table 5.1

Rigid Board and Block Insulation								
Insulation Type	**Material**	**R/in**	**Durability/Fire /Structure**	**Permeability**	**Air Barrier**	**Cost/R ($- $$$)**	**Installation Notes**	**Impact**
Straw Bale	Cereal grain straw	1.6–1.9	Low-moderate durability High fire resistance Structural	Permeable, absorptive	Yes, when plastered	$-$$	Insulation serves as enclosure material	Low recycled content, but compostable at end of life; Low toxicity, Low GHG emissions (sequesters carbon)
Fiberboard	Wood fiber	3–4	Low-moderate durability Low fire resistance Non-structural	Permeable, absorptive	Yes, for some products	$-$$$	Comparable to sheathing	High recycled content, recyclable at end of life; Low toxicity; Low GHG emissions (sequesters carbon)
Cork Board	Cork bark	4	Moderate durability Low fire resistance Non-structural	Class III, absorptive	Yes	$$$	Non-standard sizing requires full sheathing	High post-industrial recycled content, potentially recyclable at end of life (few facilities exist), compostable; Low toxicity; Low GHG emissions (sequesters carbon) →

Table 5.1 ***cont.***

Rigid Board and Block Insulation								
Insulation Type	**Material**	**R/in**	**Durability/Fire /Structure**	**Permeability**	**Air Barrier**	**Cost/R ($– $$$)**	**Installation Notes**	**Impact**
Cellular Glass	"Puffed" glass	3.4	High durability High fire resistance Structural	Class I, non-absorptive	Yes	$$$	Comparable to masonry block	High post-industrial recycled content, but not recyclable at end of life; Low toxicity; Moderate GHG emissions (embodied energy)
Mineral Board	Slag, basalt	4	High durability High fire resistance Non-structural	Permeable, less absorptive	No	$$	Select high-density board for walls, below slabs	High post-industrial recycled content, but not recyclable at end of life; Moderate toxicity (formaldehyde, irritating fibers); Moderate GHG emissions (embodied energy)
Fiberglass	Spun glass fiber with aluminum facing	4.3	Moderate durability High fire resistance Non-structural	Class I when faced, non-absorptive	Yes, when faced	$$	Standard non-structural insulated sheathing	Moderate recycled content, but not recyclable at end of life; Moderate toxicity (formaldehyde in some products, irritating fibers); Moderate GHG emissions (embodied energy)
AAC Block	"Honey-combed" cementitious concrete	1.2–2.6	High durability High fire resistance Structural	Permeable, absorptive	Yes, when parged	$$$	Comparable to concrete block	Moderate post-industrial recycled content, but not recyclable at end of life; Low toxicity; Moderate GHG emissions (embodied energy)
Polyiso/PIC, faced (yellow/ shiny)	Polyiso-cyanurate (often with Al foil or vinyl facing)	5–7	Moderate durability Low fire resistance Non-structural	Class I when faced, non-absorptive	Yes	$–$$	Standard non-structural insulated sheathing	Low-moderate recycled content, potentially recyclable at end of life (few facilities exist); High toxicity (persistent petro-chemicals and halogenated flame retardants); High GHG emissions (embodied energy)
EPS Foam (white)	Expanded polystyrene	4	Moderate durability Low fire resistance Structural (high-density types)	Class II–III, non-absorptive	Yes	$–$$	Select high-density for structural use	Low-moderate recycled content, potentially recyclable at end of life (few facilities exist); High toxicity (persistent petro-chemicals and halogenated flame retardants in most products); High GHG emissions (embodied energy) →

Table 5.1 *cont.*

Rigid Board and Block Insulation								
Insulation Type	Material	R/in	Durability/Fire /Structure	Permeability	Air Barrier	Cost/R ($- $$$)	Installation Notes	Impact
XPS Foam (pink/blue)	Extruded polystyrene	5	Moderate durability Low fire resistance Structural	Class I–II, non-absorptive	Yes	$–$$	Standard board insulation product	Low-moderate recycled content, potentially recyclable at end of life (few facilities exist); High toxicity (persistent petro-chemicals and halogenated flame retardants); Highest GHG emissions (blowing agents of most products, embodied energy)

Fig. 5.5: *Straw bales provide good thermal insulation as part of a plastered assembly, while helping to sequester carbon and reduce toxicity in the home compared to other insulation choices.*

CREDIT: ACE MCARLETON — NEW FRAMEWORKS NATURAL DESIGN/BUILD.

Batt Insulation

Batt insulation, particularly fiberglass batt, is by far the most common form of insulation in frame walls and roofs. Its use is widespread because it has many advantages: it is cheap, pre-formed yet flexible, can be easily installed into the framed structure of an assembly with few tools, and it has a paper-facing that satisfies vapor control code. However, while batt insulation — fiberglass or otherwise — is very easy to install, it is much more difficult to install well. It is rare that appropriate detailing is actually accomplished in fitting out batts to fully fill cavities and notching around wiring and other obstructions. Most often, significant void spaces are created that exacerbate convective movement in the wall, leading to significant thermal performance loss — especially with cold temperatures — and associated condensation issues.

Table 5.2

Batt Insulation								
Insulation Type	**Material**	**R/in**	**Durability/Fire /Structure**	**Moisture**	**Air Barrier**	**Cost/R ($- $$$)**	**Ease of Use**	**Impact**
Cotton	Recycled cotton fiber	3.4	Low durability Low fire resistance Non-structural	Permeable, absorptive	No	$$	Hard to cut, may slump in cavity	High recycled content, potentially recyclable at end of life (few facilities exist); Low toxicity; Low GHG emissions
Hemp	Hemp fiber	3.6	Low durability Low fire resistance Non-structural	Permeable, absorptive	No	$$$	Standard batt product	Low recycled content, but compostable at end of life; Low toxicity; Low GHG emissions (sequesters carbon)
Sheep's Wool	Sheep's wool	3.5	Low durability Moderate fire resistance Non-structural	Permeable, absorptive	No	$$$	Hard to cut	Low recycled content, but compostable at end of life; Low toxicity; Low GHG emissions
Mineral Wool	Slag, basalt	4	Moderate durability High fire resistance Non-structural	Permeable, non-absorptive	No	$$	Hard to cut, fibers are irritating	High post-industrial recycled content, but not recyclable at end of life; Low toxicity (irritating fibers); Moderate GHG emissions (embodied energy)
Fiberglass	Spun glass fiber	3.5–3.8	Moderate durability Moderate fire resistance Non-structural	Permeable, non-absorptive	No	$	Rarely installed correctly, fibers are irritating	Moderate recycled content, but not recyclable at end of life; Moderate toxicity (formaldehyde in some products, irritating fibers); Moderate GHG emissions (embodied energy)

Spray-applied Insulation

Whereas cellulose and fiberglass spray-applied insulations are similar to their cavity-fill counterparts (and discussed below), spray foam is nearly a category unto itself, as there are different types with different characteristics. Spray foams can be used just as air-sealing and spot insulation (simple cans from the hardware store), but as insulation they are applied as a two-part chemical mix, which usually needs to be done by professional installers. The flexible application of a product that can act as an air barrier, insulation, water control, and (in the

Table 5.3

Spray-applied Insulation								
Insulation Type	**Material**	**R/in**	**Durability/Fire /Structure**	**Moisture**	**Air Barrier**	**Cost/R ($–$$$)**	**Ease of Use**	**Impact**
Cellulose	Shredded paper, borate	3.5	Moderate durability Moderate fire resistance Non-structural	Permeable, absorptive	No, but reduces airflow	$$	Requires special equipment, drying time	High recycled content, recyclable/compostable at end of life; Low toxicity; Low GHG emissions
Fiberglass	Shredded glass fiber, acrylic	3.5	Moderate durability Moderate fire resistance Non-structural	Permeable, non-absorptive	No	$$	Requires special equipment	Moderate recycled content, but not recyclable at end of life; Low toxicity; Moderate GHG emissions (embodied energy)
Icyenene	Formaldehyde foam	3.7	Moderate durability Low fire resistance Non-structural	Class III, non-absorptive	Yes	$$$	Requires special equipment, hazardous during installation	No recycled content, not recyclable at end of life; High toxicity (persistent petro-chemicals and halogenated flame retardants); High GHG emissions (embodied energy)
Open-cell Foam	Polyurethane foam	3.6	Moderate durability Low fire resistance Non-structural	Class III, non-absorptive	Yes	$$$	Requires special equipment, hazardous during installation	No recycled content, not recyclable at end of life; High toxicity (persistent petro-chemicals and halogenated flame retardants); High GHG emissions (embodied energy)
Closed-cell Foam	Polyurethane foam	6.5	Moderate durability Low fire resistance Non-structural	Class I–II, non-absorptive	Yes	$$$	Requires special equipment, hazardous during installation	No recycled content, not recyclable at end of life; High toxicity (persistent petro-chemicals and halogenated flame retardants); Highest GHG emissions (blowing agents of most products, embodied energy)

case of closed-cell) vapor barrier has made these products very popular in the green building market. Foams only work when they are applied well, however, which involves conducting on-site chemistry. If substrate conditions are not within guidelines (temperature, cleanliness, moisture); equipment is not adjusted properly; and the applicator is not alert, well-trained, and following manufacturer's instructions, there can be major problems with the quality of the installation. Spray foams feature the same range of ecological, carbon, and toxic footprints and fire risk as their board cousins to varying degrees — newer closed-cell foams using lower GWP blowing agents are approaching emissions levels comparable to those of rigid mineral board, for example.[3]

Cavity-fill Insulation

One of the more common cavity-fill insulations is cellulose, which can be installed "loose" into an attic where it can settle a couple of inches, "dense-packed" under pressure into a closed-cavity to a high density to overcome settling, or "damp-sprayed" with a mist of water into an open cavity. They are made primarily from waste paper, with small amounts of the mineral borate (or in the case of poorer quality products, ammonia compounds) formulated into the paper to make it repellent to fire, insects, and mold. Of particular benefit is its ability to fully fill a cavity; because it is dense, it is very effective in mitigating convective loops and leaks. Note, though, that dense-pack applications require specialized equipment and trained installers to avoid

Table 5.4

Cavity-fill Insulations							
Insulation Type	Material	R/in	Moisture	Air Barrier	Cost/R ($– $$$)	Ease of Use	Impact
Bonded Cellulose	Clay or lime binder, natural cellulose (i.e. wood, hemp, straw)	1.2–2	Permeable, absorptive	No, but reduces air flow	$–$$	Formulation and installation procedures vary widely	Low recycled content, but compostable at end of life; Low toxicity; Low GHG emissions (sequesters carbon)
Cellulose, loose	Shredded paper, borate	3.8	Permeable, absorptive	No, but reduces air flow	$	Simple to install, requires specialized equipment	High recycled content, recyclable/compostable at end of life; Low toxicity; Low GHG emissions
Cellulose, dense-pack	Shredded paper, borate	3.5	Permeable, absorptive	No, but reduces air flow	$–$$	Requires specialized equipment and training	High recycled content, recyclable/compostable at end of life; Low toxicity; Low GHG emissions
Perlite	Puffed mineral	2.5–3.5	Permeable, less absorptive	No	$–$$	Simple to install in cavities	Low recycled content, inert for safe end of life disposal; Low toxicity; Moderate GHG emissions (embodied energy)
Blown-in Fiberglass	Shredded glass fiber	3.5	Permeable, non-absorptive	No	$	Simple to install, requires specialized equipment	Moderate recycled content, but not recyclable at end of life; Low toxicity; Moderate GHG emissions (embodied energy)

settling and associated thermal and durability problems. Cellulose is known for being able to hold and distribute moisture better than most insulations.

Loose or densely-blown fiberglass is a common alternative to cellulose. Loose fiberglass does not offer the same hygroscopic benefits as cellulose, and it is less dense, which can lead to convective air cycling and thermal loss depending on the application. Newer spray-applied systems are similar to damp-spray cellulose, relying on an acrylic binder to stabilize the installation.

There is a whole, relatively unknown, category of bonded cellulosic fill insulations available to builders; these are made of natural materials bound with either clay or lime. Straw, wood chips, and hemp hurds are the most common materials, the former two commonly bonded with clay, and the latter with lime to create hempcrete. As with the other fill materials, they excel in being able to fully fill a cavity, providing maximum insulation and minimizing convective losses. Formulations vary, as they are often manufactured on site, and they require a skilled practitioner to ensure the quality of the mix and installation. There are built-in moisture considerations with these materials, as noted in the previous chapter, but once dry they offer terrific moisture storage capacity, with durability enhanced by the clay and lime binders.

Finally, mineral-based fill insulations date back hundreds of years. Most common is perlite, a "puffed" stone that expands when heat-processed, creating microscopic glass bubbles. The most common application is as a fill in masonry block or cavity construction, although other applications range widely, including use as an aggregate below slabs and formulated into insulation boards.

Fig. 5.6: *Dense-pack cellulose insulation offers the advantages of fully filling the cavity and attenuating air flow, and can be used in combination with other insulation, such as straw bale, as shown here.* Credit: Ace McArleton — New Frameworks Natural Design/Build.

Windows: Part of the Thermal Enclosure

Windows do many things for us — provide views, solar energy, light, air, and more. They are also inherent compromises in otherwise robust enclosures, from both a moisture and a thermal perspective (with the exception of windows that provide net thermal gain or daylighting as part of a passive solar strategy). There are a few particular considerations in selecting windows as part of our thermal enclosure (see the sidebar "Criteria for Selecting Windows" for more on window selection). Note that window ratings are given as U-values rather than R-values — simply divide 1 by the U-value to get the R-value.

Fig. 5.7

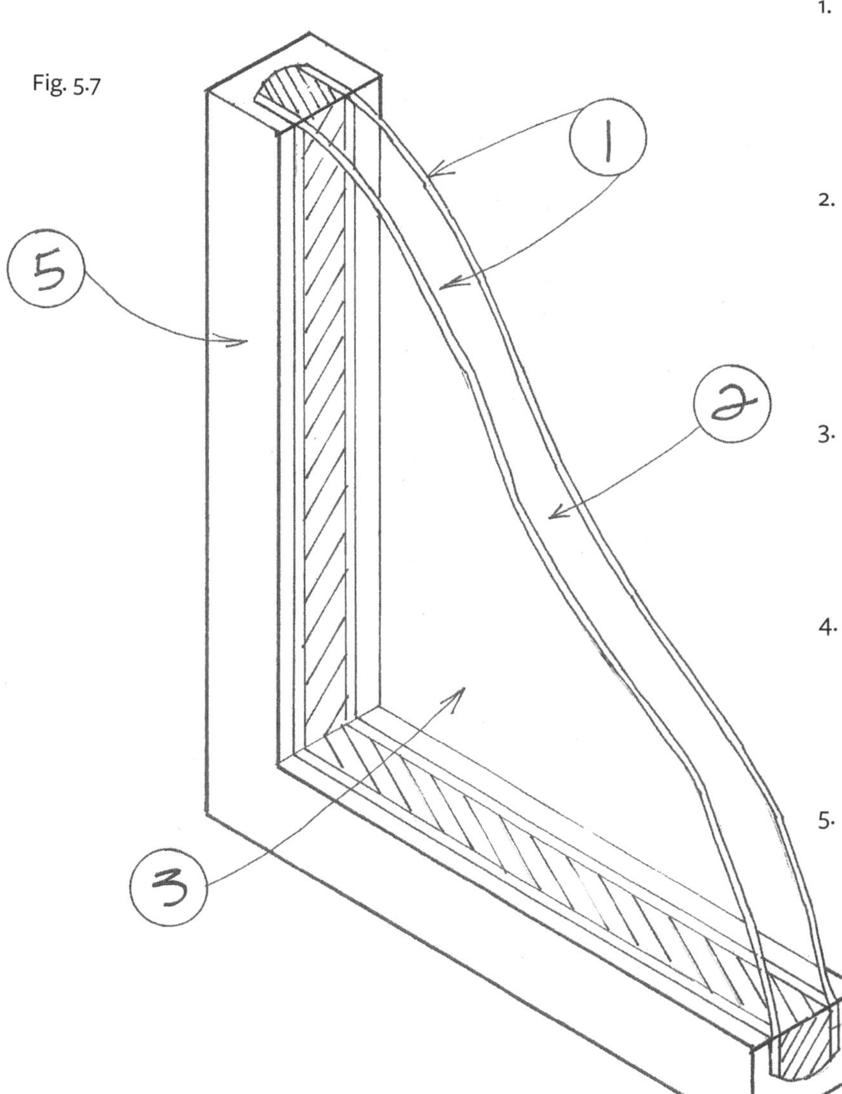

1. *Multiple Panes:* With more panes come more air spaces, which provide insulation and sound attenuation benefits. While double-pane windows have long been the standard, triple-pane units are much more accessible now.
2. *Gas Fills:* Windows filled with inert "noble" gases, such as argon or krypton, between the panes have higher insulation values than those filled with air. These gases are odorless, colorless, and nontoxic, but will leak and be replaced with air over time.
3. *Low-e Coatings:* ("e" stands for "emissivity") These can be applied to the glazing, reducing radiant heat transfer and further boosting the thermal performance. They also reduce UV transmission, reducing fading of interior finishes.
4. *Warm-edge Spacers:* In energy-efficient windows, the spacers that separate the window panes from each other should be low-conductivity to reduce conduction and minimize condensation on the glass.
5. *Frames:* Efficient window frames are often made of either fiberglass, PVC, or wood composite. Durability, conductivity, cost, and ecological impact are some of the driving factors here.

Table 5.5

Frame Material	Advantages	Disadvantages
PVC/uPVC	• very inexpensive • relatively durable when protected from UV • strong material that can be built to incorporate thermal breaks	• PVC is a particularly toxic plastic (on Living Building Challenge Red List) • persists in the environment for a very long time • susceptible to UV damage if not protected/treated
Fiberglass	• significantly less ecologically damaging than PVC • strong material that can be built to incorporate thermal breaks • relatively durable when protected from UV • has almost identical thermal expansion and contraction behavior as the glazing itself, promoting longevity of window.	• more expensive than PVC • susceptible to UV damage if not protected/treated
Wood and Wood Composite	• least ecological impact (especially solid wood) • relatively durable when protected from UV • attractive natural wood finish	• most expensive option • solid wood is more conductive, and harder to incorporate thermal breaks into the frame (frame is often solid wood) • susceptible to UV damage if not protected/treated

Fig. 5.8: *The fiberglass-frame window like the one being installed here is moderately-priced, strong, and durable, and is more environmentally-friendly than PVC.* CREDIT: ACE MCARLETON — NEW FRAMEWORKS NATURAL DESIGN/BUILD.

Criteria for Selecting Windows: The NFRC Label

In the United States, all windows are rated for energy efficiency by the National Fenestration Rating Council (NFRC) and receive a label featuring 3 to 5 ratings (three are mandatory, two are optional).[4] These ratings allow us to easily compare units. Note that whole units are evaluated here — frames, spacers, and glazing. U-values for "Center of Glass (CoG)" only look at performance through the middle of the glazing; they are not comparable to the whole-unit evaluation of the NFRC label. Note that in both the US and Canada, the North American Fenestration Standard (NAFS), which addresses structural performance, water penetration, and air tightness, is referenced in codes, and differ between the two countries; see the "Resources" section for further information. Here is a quick breakdown of the criteria:

- **U-value:** As discussed earlier, U-value is a measurement of the conductivity of the unit in Btu/ft^2·°F·hr (kW/m^2·°C). The higher the U-value, the more heat transfer. Divide 1 by the U-value to get the R-value. The higher the R-value, the greater the resistance to heat flow.
- **Solar Heat Gain Coefficient (SHGC):** This is the fraction of solar radiant heat that can transfer through the unit; it is expressed as a number between 0 and 1. The higher the number, the more solar heat can transfer through the unit. This is helpful to optimize for passive solar gain.
- **Visible Transmittance (VT):** This is a fraction of the visible light that can transfer through the unit, and is expressed as a number between 0 and 1. The higher the number, the more visible light can transmit through the unit.
- **Air Leakage (AL):** An optional measurement, this value represents the amount of air that can leak through the unit, expressed as cubic feet per minute per square foot (CFM/sq. ft.) at a given test pressure. The lower the CFM/sq. ft., the tighter the window.
- **Condensation Resistance (CR):** Also an optional measurement, this value represents how well a window will resist the formation of condensation on the inside surface of the unit. Represented as a scale from 1 to 100, CR is based on the interior surface temperature at a given outdoor air temperature and wind speed. The higher the number, the more resistant the unit is to condensation.

In order to decide how to specify your windows, it's important to place them in the context of the overall thermal performance goals, as well as your particular climate and site. Optimizing U-value is worthwhile in any climate, and optimizing cost versus performance will be informed by your goals. Tuning the SHGC is important, and this is where climate counts. In hot climates, you will generally reduce SHGC in the entire building to limit cooling loads. In cold climates in the Northern Hemisphere, SHGC is less relevant on northern walls, but will be critical on the south for passive solar-oriented homes (the reverse is true in the Southern Hemisphere). A high SHGC on eastern walls is helpful to warm the house up in the morning, but can lead to overheating on the west, as the building is likely already up to temperature by later in the day. These specs must both balance against VT to ensure the quality of light. Martin Holladay, editor at GreenBuildingAdvisor.com, recommends these ranges (all for whole-window values)[5]:

Hot climates, moderate performance:

- U-0.30 or less (lower is better)
- SHGC-0.28 to 0.37 (lower is better)

Cold climates, moderate performance:

- U-0.30 to 0.39 (lower is better)
- SHGC-0.42 to 0.55 (for south/east orientation — higher is better)

High performance windows:

- U-0.19 to 0.26 (lower is better)
- SHGC-0.33 to 0.47 (for south/east orientation in cold climates — higher is better)

Thermal Performance Goals: How Much Is Enough?

Now that we've identified what makes up the thermal enclosure and some of the materials used in its composition, we must decide how much insulation we need, and how airtight our building needs to be. Let's look at the questions you might ask in this regard, and how the answers might influence your R-value and airtightness specifications for your assemblies.

What is my climate?

The more extreme temperatures experienced in your climate, the more insulation you will need to avoid discomfort or excessive energy consumption. *Heating degree days (HDD)* or *cooling degree days (CDD)* are the metrics used to quantify and evaluate this. HDDs and CDDs are measures of how much (in degrees) and for how long (in days) the outside air temperature was either higher (for CDD) or lower (for HDD) than an indoor air temperature baseline (commonly 65°F (18°C) for HDD and 70°F (21°C) for CDD).[6] For energy modeling, the standard is

Table 5.6

Zone Number	International Climate Zone Definitions[7]	
	Thermal Criteria	
	IP Units	*SI Units*
1	9000 ≤ CDD50°F	5000 ≤ CDD10°C
2	6300 ≤ CDD50°F ≤ 9000	3500 ≤ CDD10°C ≤ 5000
3A and 3B	4500 ≤ CDD50°F ≤ 6300 and HDD65°F ≤ 5400	2500 ≤ CDD10°C ≤ 3500 and HDD18°C ≤ 3000
4A and 4B	CDD50°F ≤ 4500 and HDD65°F ≤ 5400	CDD10°C ≤ 2500 and HDD18°C ≤ 3000
3C	HDD65°F ≤ 3600	HDD18°C ≤ 2000
4C	3600 ≤ HDD65°F ≤ 5400	2000 ≤ HDD18°C ≤ 3000
5	5400 ≤ HDD65°F ≤ 7200	3000 ≤ HDD18°C ≤ 4000
6	7200 ≤ HDD65°F ≤ 9000	4000 ≤ HDD18°C ≤ 5000
7	9000 ≤ HDD65°F ≤ 12600	5000 ≤ HDD18°C ≤ 7000
8	12600 ≤ HDD65°F	7000 ≤ HDD18°C

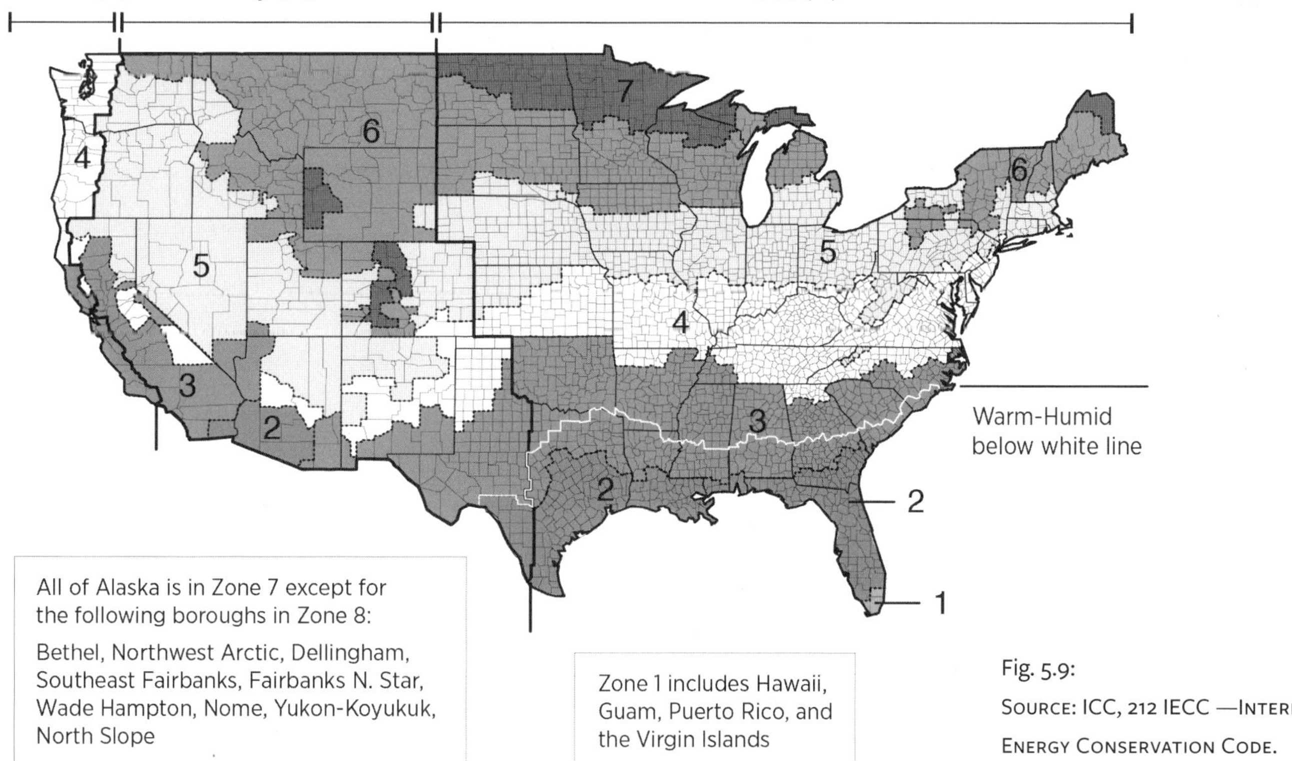

Fig. 5.9: Source: ICC, 212 IECC —International Energy Conservation Code.

an annual average based on 30 years of historical averages. Another climate-related impact is the amount of solar radiation, which depends on both latitude and average seasonal cloud cover (along with building siting and orientation choices).

What are my goals and intentions?

We opened the book in Chapter 1 with an evaluation of goals and intentions to help guide your formulation of strategies, and we return to those considerations here. There are many different reasons to specify a high-quality enclosure, including comfort and indoor health, building durability, low operating costs, ecological sensitivity/climate response, and resilience. Having a clear sense of these priorities is critical to making good decisions. One of the most basic decisions you need to make is whether to optimize enclosure quality for economic or for ecological gain. If the former, you will be evaluating reduced operating costs and monthly cash-flow against up-front and financing costs. If the latter, you will be evaluating embodied and operational carbon emissions based on different design options, with a consideration of the value (rather than up-front cash outlay) of these enhancements.

What is my building profile and framing style?

There are many variables to consider, from material selection to the quality of the thermal control layer and air barrier to the cost of installation, and your overall goals will guide these decisions. In the design phase, apply the advantages and disadvantages listed in Table 5.7 to a few different major assembly types to help guide your decision-making.

Table 5.7

Assembly Type	Description	Advantages	Disadvantages
Foundation			
Slab-on-grade	Earth-coupled slab at grade level (first floor)	• Less excavation, concrete (therefore cost, impact) • Slab provides thermal mass, hydronic heat potential, durable floor • Different styles; design flexibility	• Less professional familiarity in cold climates (perceived risk) • More complicated/expensive on sloped sites • Harder to modify utilities buried beneath slab in future • Less bottom-of-wall protection
Basement	Extra deep frost walls create below-grade level	• Standard and familiar, esp. in cold climates • Durable and proven, low risk • Can provide inexpensive additional space	• Often expensive to provide adequate thermal and moisture control, esp. in cold/wet climates • Frequently executed poorly, creating cold and dank spaces • Requires use of high-impact materials, additional excavation
Piers	Structural columns run below frost depth supporting floor deck or structure	• Standard and familiar in all regions • Durable and proven, low risk • Least expensive option	• Provides no enclosure at base of building; can lead to heat loss • Complicates running water into building in freezing climates • Requires careful detailing to avoid individual post heaving/settling
Walls/Structure			
Stud-frame	Standard residential frame, optional wrap with insulation	• Inexpensive, common • Easy to build • Lots of design options	• Harder to get high-R w/o foam • Vulnerable in cold climates if not well-protected →

Table 5.7 *cont.*

Assembly Type	Description	Advantages	Disadvantages
Walls/Structure			
Double-stud	Two stud-frame walls in assembly, providing a deep insulation cavity	• Simple, common construction • Eco-friendly materials • High-R, durable wall	• Higher labor, more expensive than single stud wall • Design details potentially more complex
Timber Frame	Post-and-beam structural frame, no assembly enclosure	• Beautiful, historic • Very durable • Eco-friendly material (depends on wood source, treatment)	• Doesn't provide enclosure • More expensive • Specialized labor and/or equipment
SIP (Structural Insulated Panel)	Pre-fabricated structural board insulation-core assembly panels for roofs and walls	• Can be cheaper than custom assembly • Quick installation • High-R, reduced thermal bridging	• Often relies on high-impact foam (some alternatives available) • Specialized labor and/or equipment • Limited design flexibility
ICF (Insulated Concrete Form)	Leave-in-place concrete-core forms with interior and exterior insulation	• Very durable and structural, above- and below-grade (foundation) • Built-in insulation, air-tight • Quick installation	• Often relies on high-impact foam (some alternatives available) • More expensive • Specialized labor and/or equipment
Roof			
Attic	Unconditioned space above insulated ceiling and below roof	• Cheapest to insulate and air-seal to high levels • Very durable, allows drying and roof ventilation • Allows easy utility runs below insulation with future access	• May raise profile of building in some designs • Interior attic hatches are often leaky and poorly-insulated • Attic space not very useful (hard to access, very hot/cold)
Vented Cathedral	Insulated roof structural assembly with roof sheathing ventilation	• Allows conditioned use of all space below roof structure • Venting allows drying potential, reduces ice damming • May reduce profile of building in some designs	• Harder to get high insulation levels and avoid thermal bridging • Venting may reduce insulation space or add extra layers and cost • Venting often done improperly
Unvented Cathedral	Insulated roof structural assembly without roof sheathing ventilation	• Allows conditioned use of all space below roof structure • Insulation fully fills cavity, no additional layers • May reduce profile of building in some designs	• Harder to get high insulation levels and avoid thermal bridging • Higher risk in cold climates, can only dry to inside • Greater potential for ice damming, condensation damage

What codes, standards, or guidelines are governing my project?

Depending on where you live, you will have code minimums for the quality of your thermal enclosure, and possibly some prescriptive measures as well. This is a crucial place to start, as your building codes are the "law of the land" and must be followed. Beyond that, there are many different standards and rating systems you may find appropriate for your building, from state/provincial energy programs to federal energy programs (i.e., EnergyStar for Homes) to prescriptive-based green building programs (i.e., LEED) to performance-based energy programs (Passive House). There are also some great standards that are not tied to certification programs, but rather just decent "rules of thumb" to follow for any given region. One popular one is the "1-5-10-20-40-60" enclosure, specifying 1 ACH50 airtightness standard, R-5 windows, R-10 below the unheated slab, R-20 below-grade walls, R-40 above-grade walls, R-60 ceiling (note, however, that many designers prefer R-20 below the slab in cold climates).[8]

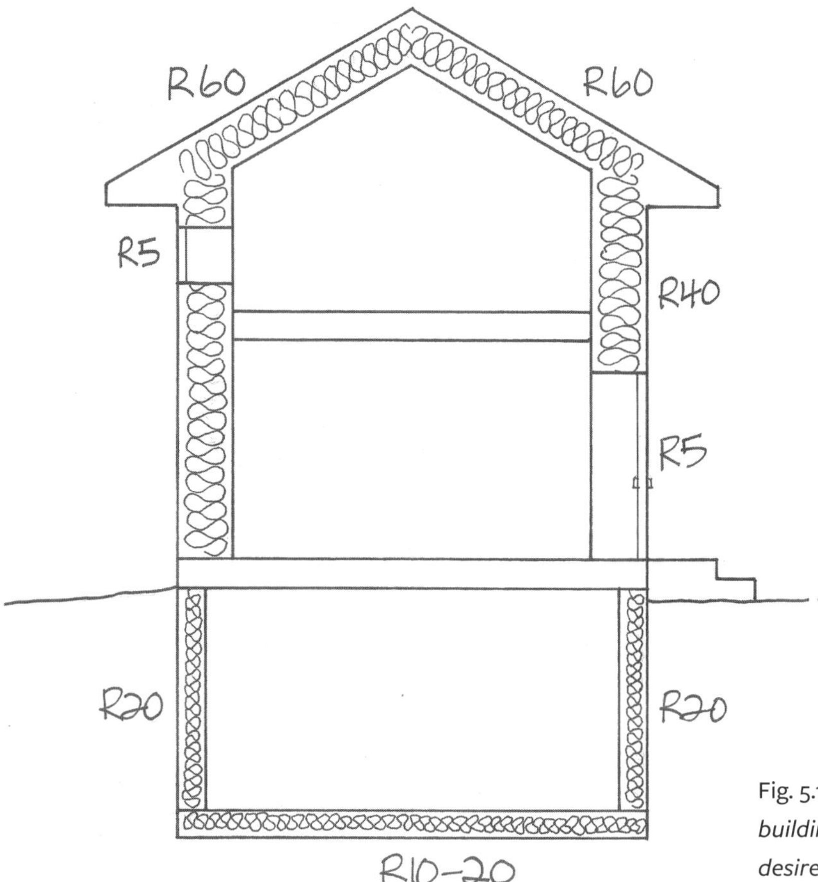

Fig. 5.10: *The thermal performance of all components of a building's enclosure must be well-matched to achieve the desired goals.*

Table 5.8

Program Name	Type of Program	Prescriptive or Performance	Climate-Sensitive	Cost of Implementation	Level of Performance
IECC	Code	Both	Yes	baseline	Baseline — minimum
EnergyStar	Certification	Prescriptive	No	$	Basic energy improvements
LEED	Rating	Both	Yes	$$–$$$	Multi-tier — good to best
PassiveHouse — PHIUS	Certification	Performance	Yes	$$–$$$	Highest-tier performance
PassiveHouse — PHI	Certification	Performance	No	$$–$$$	Highest-tier performance
"Pretty Good House"	Guideline	Performance	No	$$	Cost-optimized performance
1-5-10-20-40-60	Guideline	Performance	No	$$	Cost-optimized performance

How Airtight?

In Chapter 4 we looked closely at how air barriers control humidity. Therefore, if we are increasing our insulation levels to hit performance goals, we need to also ensure that our air control layer is brought to a similar level of quality, both to ensure the performance of the insulation (for air-permeable insulation) and the overall thermal performance of the enclosure, and also to maintain durability in our assemblies. As insulation levels go up, drying potential goes down, and airtightness becomes increasingly critical to minimize the amount of humidity in the assembly. "How much" relies on metrics. As explained in Chapter 4, we use a blower-door test to assess the amount of air leakage in a building, described as cubic feet per minute at the testing standard of 50 pascals pressure difference between the inside and outside of the building (CFM50). We need to put this value into a metric relative to our particular building, in relation to either the volume or surface area. The volume metric is *air changes per hour at 50 pascals* (ACH50, or the volume of the house in cu. ft. × 60 minutes per hour/CFM), and the surface area metric is *CFM50/ext. sq. ft.*, where the exterior square footage is generally the above-grade shell (since very little air leaks through below-grade assemblies). So how tight is tight enough? Some reference points appear in Table 5.9.

How tight is too tight? This is a matter of great debate.

Limiting air leakage limits thermal losses, and high levels of energy efficiency cannot be achieved with a leaky building. If a project goal is the best possible efficiency, that necessarily involves a strategy of best possible airtightness.

Many experts argue that for the sake of the enclosure, the tighter the better for both thermal and moisture considerations.[9] Less air leakage introduces less moisture into the enclosure, so projects aiming for the best possible durability will likewise need to strategize for best possible airtightness.

Best possible efficiency and durability goals don't suit all projects. There is general agreement that airtightness in the range of 1–2 ACH50 will suffice to avoid humidity issues and discomfort in cold climates for many common enclosure types.[10]

It must be stated that appropriate mechanical ventilation is absolutely critical to ensure indoor air quality in all cases, a topic we will cover in Chapter 7.

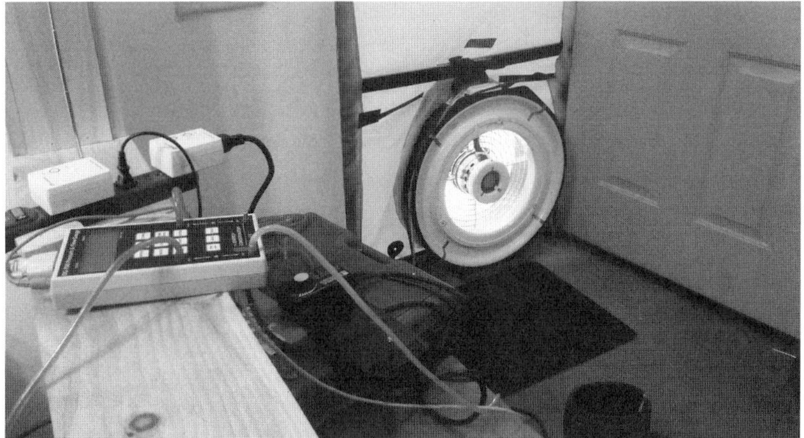

Fig. 5.11: *The blower door is a critical piece of diagnostic equipment for establishing baseline on existing buildings, conducting mid-stream quality control, directing active air-sealing work, establishing final airtightness level, and diagnosing building problems.*

CREDIT: JACOB DEVA RACUSIN — NEW FRAMEWORKS NATURAL DESIGN/BUILD.

Table 5.9

Airtightness			
Standard	**Requirement**	**Difficulty**	**Performance**
International Residential Code 2012	≤ 5 ACH50 (Climate Zones 1–2), ≤ 3 ACH50 (Climate Zones 3–8)	Easy	Good — provides sufficient performance in most cases
General High-Performance Recommendations	≤ 1–2 ACH50	Easy-Moderate	Better — optimizes thermal and moisture control in all climates
PassiveHouse (PHIUS, PHI)	≤ 0.6 ACH50	Moderate-Hard	Best — managing pressure balance and indoor RH becomes a consideration

Thermal Enclosure Design Using Energy Modeling

There are a number of different ways to design the thermal enclosure of our buildings. If you seek to go beyond the prescriptions laid out by building codes, certifications, or rule-of-thumb approaches, or if you want the ability to track the thermal impact of changes made during the design process, energy modeling is important. Energy models can be used to evaluate the building's performance from many perspectives, so they should be selected to answer the specific questions you have.

Learning how to use energy modeling tools (or working with an experienced professional) is an incredibly effective and powerful way to understand how different design decisions impact the performance of a building. It is a critical component of any high-performance building design, and need not be a complex and inaccessible technology. Quite the contrary, these programs can make seemingly arbitrary and mystifying decisions — for example, how much insulation to put in the attic, or the effects of two different foundation systems on energy consumption — much more tangible, data-driven, and predictable; a good modeler with a good tool will be able to predict the outcomes of different choices to a reasonable degree of accuracy. Learning when to use modeling tools is also important; buildings that are modeled early in the design phase can still be modified to optimize performance based on the modeled results, whereas models run at the end of the design process are generally done too late to impact design decisions, and serve as a benchmark rather than a design tool.

You need to select the right model for the job. Here are some of the features you might look for in modeling software:

- *Annual Energy Loads:* most basic level, to evaluate how enclosure design affects energy performance.
- *Peak Energy Loads:* calculates energy loads at hottest and coldest times of the year to inform mechanical system sizing; may take into account mechanical system efficiencies, internal and solar gains, and peak domestic hot water (DHW) demand for hot water system sizing.
- *Room-by-Room Analysis:* calculates energy loads on a per-room basis to inform mechanical system distribution and design.
- *Cooling, Latent Energy, Ventilation Loads:* critical for cooling-dominated systems; calculates cooling loads, or energy impact, of humidity and the energy impact of ventilation based on the equipment sizing and efficiency.
- *Total Building Energy/Renewable Energy:* includes inputs for DHW, non-HVAC electric loads, primary energy, and renewable energy sizing to achieve net zero energy or carbon neutrality.
- *Program Certification:* modeling results can be used to apply for energy program certification.

More complicated programs may offer more precision, but not necessarily more accuracy — more error in more inputs can lead to worse results than simpler models! However, they can offer more control and features, which may be necessary for making decisions based on performance. It is important to have a qualified person running the model to ensure that you receive accurate information — the old adage in modeling is "Garbage In, Garbage Out" (GIGO).

In addition to spreadsheet and database energy models, 3D modeling is possible now, with plug-ins and features for programs such as SketchUp and Revit allowing for integrated energy modeling in the design phase; the ability to draw in layers for the air barrier and insulation

Table 5.10[11]

Program	Cost	Difficulty	Annual Energy?	Peak Energy/ System Sizing?	Room-By-Room/ System Design?	Cooling And Ventilation Loads?	Total Building Energy and PV?	Program Certification?
EnergyPlus, eQUEST	Free	Easy/ Moderate	Yes	Yes	Yes	Yes	Yes	No
REM/Rate	Provided by HERS rater	Moderate/ Hard	Yes	Yes	No	Yes	Yes	HERS, EnergyStar, LEED, IECC
HOT2000	Provided by rater/ professional	Moderate/ Hard	Yes	Yes	No	Yes	Yes	R-2000, EnerGuard, EnergyStar
Manuals J & D	Free– >$1k	Easy	No	Yes	Yes	Yes	No	ACCA
PHPP	$100–$300	Hard	Yes	Yes	No	Yes	Yes	PassiveHouse

also aid evaluation of build-ability or potential thermal breaks — all in 3D. Further, these programs allow us to conduct solar and daylighting analysis to inform the effects of siting, building form, and layout on energy performance.

While a good energy model can inform a quality design, it won't guarantee a quality result. Assessing quality control during construction through oversight, inspections, and diagnostics (such as running a blower-door test or using an IR camera) is critical to ensure the building will be built to the standard represented in the model. Tracking the actual energy consumption after the work has been done and the building is in use will provide a valuable validation of the model, which can be used to improve future models or even troubleshoot potential problems with the energy performance.

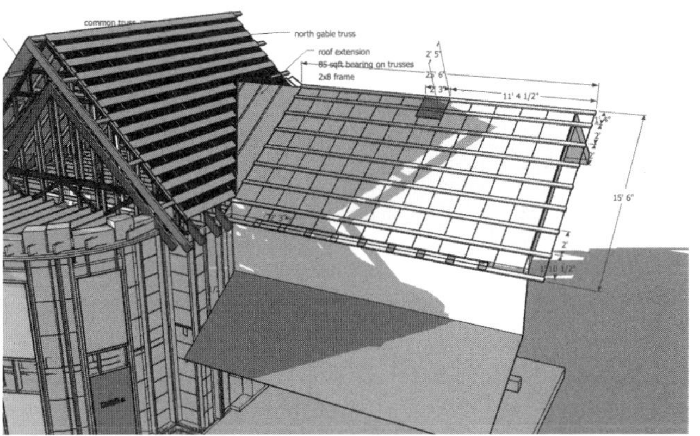

Fig. 5.12: *3D modeling technology offers a wealth of design control to maximize building performance, from modeling solar exposure to working out complex enclosure transitions.*
CREDIT: BEN GRAHAM — NEW FRAMEWORKS NATURAL DESIGN/BUILD.

Chapter 6

Examples of Building Assemblies

Foundations

Foundations can be the source of many building woes, especially moisture-related problems. Building science can guide us in the design and construction of well-performing foundations and the remediation of foundation problems. While we cannot be comprehensive in our exploration of the many options and conditions surrounding foundation detailing, we will look at a few basic foundation types, identifying the locations of the different layers and describing their characteristics, advantages and disadvantages. It is strongly encouraged that the reader consult other resources to ensure proper detailing.

Dealing with Surface and Groundwater

Regardless of the foundation type, there are some standard drainage techniques involving the management of bulk water that need to be understood even before we start talking about layers of the enclosure. Some of the standard procedures/practices are shown in Figure 6.1.

1. Gutters with downspouts reduce splashback and facilitate redirection of the water safely away from the building and allow for rainwater catchment.
2. Positive grading away from perimeter of the foundation, at a 5% pitch away from the building for ten feet, or to a swale or other positive drain. Installation of a "cap" of water-resistant material to shed water away from foundation and diffusing surface material such as pea stone to reduce splash-back if gutters are not present.
3. Back-fill the foundation with a drainable material (such as bank-run gravel) to the footing drain to facilitate drainage.
4. Drainage at the footing level, wrapped in filter fabric and permeable fill, and drained to daylight, a dry well, or a sump pump.
5. Granular drainage plane below slab (such as clean crushed stone — no fines).

Fig. 6.1

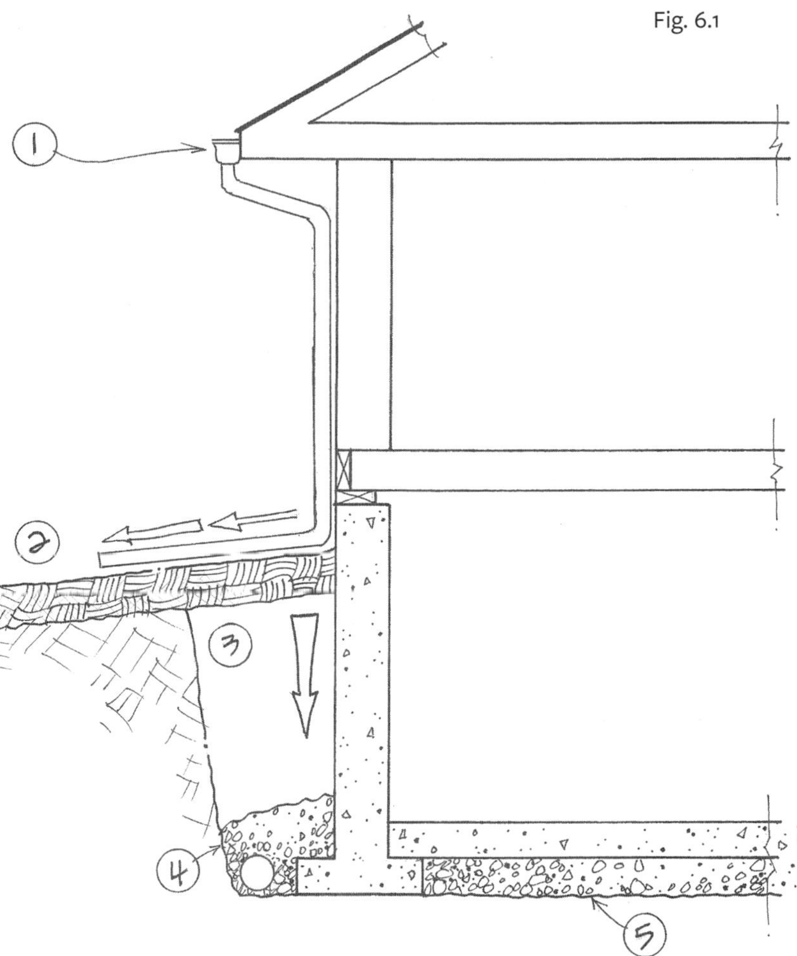

Basement with Exterior Insulation

This foundation style is common in many regions in North America, and it has a mixed history, with plenty of examples of successes and failures. By paying close attention to water control and thermal control layers, the chances of failure can be diminished. It should be noted that any type of deep foundation is inappropriate in areas with a consistently high water table or a history of flooding; the commonly used features and practices shown in Figure 6.2 provide water control, but not water prevention in high water-risk environments.

Advantages

- Exterior insulation keeps the masonry wall warm and reduces condensation risk.
- Below-grade insulation is protected from damage (i.e., UV, fire, impact) as access is greatly reduced.
- Hydrophobic insulation helps protect foundation wall.
- Foundation can be brought above grade to help protect vulnerable bottom-of-wall details from surface water, splashback, etc.
- Large amount of mass inside the thermal enclosure directly coupled to the basement environment can provide thermal benefits.
- Standard design uses common materials and trades practices.
- Basements may provide economical and/or practical space, whether finished or used as storage, mechanical, or utility space.

Disadvantages

- Above-grade insulation is vulnerable to damage (i.e., UV, insects, fire, weed-whackers) unless very well-protected; many common trowel-applied protection layers fail easily.
- Wet-service location requires more ecologically damaging materials (i.e., foam, mineral board), which may conflict with project values and goals.
- Thermal bridge-free construction around the footings may be more complicated.

1. Rigid insulation suitable for direct contact with soil (i.e., mineral board, fiberglass board, high-density EPS or XPS foam). Note that this layer acts as a water-control layer (and as vapor control, if the material itself is vapor-impermeable) as well as thermal control. Termite control may be a concern for subgrade insulation in certain regions.
2. Exterior treatments to protect above-grade insulation from insects, UV degradation, and erosion; these are installed beneath flashing, between above-grade wall and foundation.
3. Capillary break (i.e., structural gasket, liquid-applied sealant) to ensure liquid moisture in the foundation wall cannot wick up into sensitive parts of the wall assembly. Gaskets are a great solution, as they can also serve to control air through this junction as part of the air-control layer.
4. Sealant to provide continuity of air-control layer: a) from foundation wall to mud sill (if not provided by capillary break), b) from mud sill plate to band/rim joist, c) from band/rim joist to subfloor, d) from subfloor to wall base plate, e) from wall base plate to interior air barrier materials, and/or f) from wall base plate to exterior air barrier materials.
5. Band-joist insulation (i.e., cellulose, batt, spray foam).
6. Water-control layer placed directly against the foundation wall. This redundancy backs up the water control role of the insulation and provides vapor control in the instance of a vapor-permeable insulation such as mineral board. Materials for this layer include peel-and-stick membranes or liquid-applied moisture barriers.
7. The concrete slab and walls serve as the air barrier, provided that all penetrations and transitions, such as plumbing and electrical stubs, sump pump housings, and doors and windows, are installed to be airtight.
8. Capillary break — liquid-applied or sheet membrane — between wall and footing is required when footing is in direct contact with soil.
9. Concrete footing below frost depth (note thermal bridge at uninsulated footing without interior wall insulation).
10. Water/vapor control layer below the slab is a sub-slab-grade polyethylene membrane, and may act as a radon barrier. All seams must be sealed, and membrane must be permanently sealed to foundation walls.
11. Sub-slab rigid insulation (i.e., high-density mineral board, high-density EPS or XPS foam). Slab insulation levels must be higher for slabs fitted with radiant heating systems. Note potential thermal bridge through footing if footing is not insulated — may not be suitable in cold climates. Termite control may be a concern for subgrade insulation in certain regions.
12. Granular capillary break/drainage pad (i.e., crushed stone or gravel) reduces bulk moisture wicking into slab and facilitates interior drainage and soil gas mitigation systems.

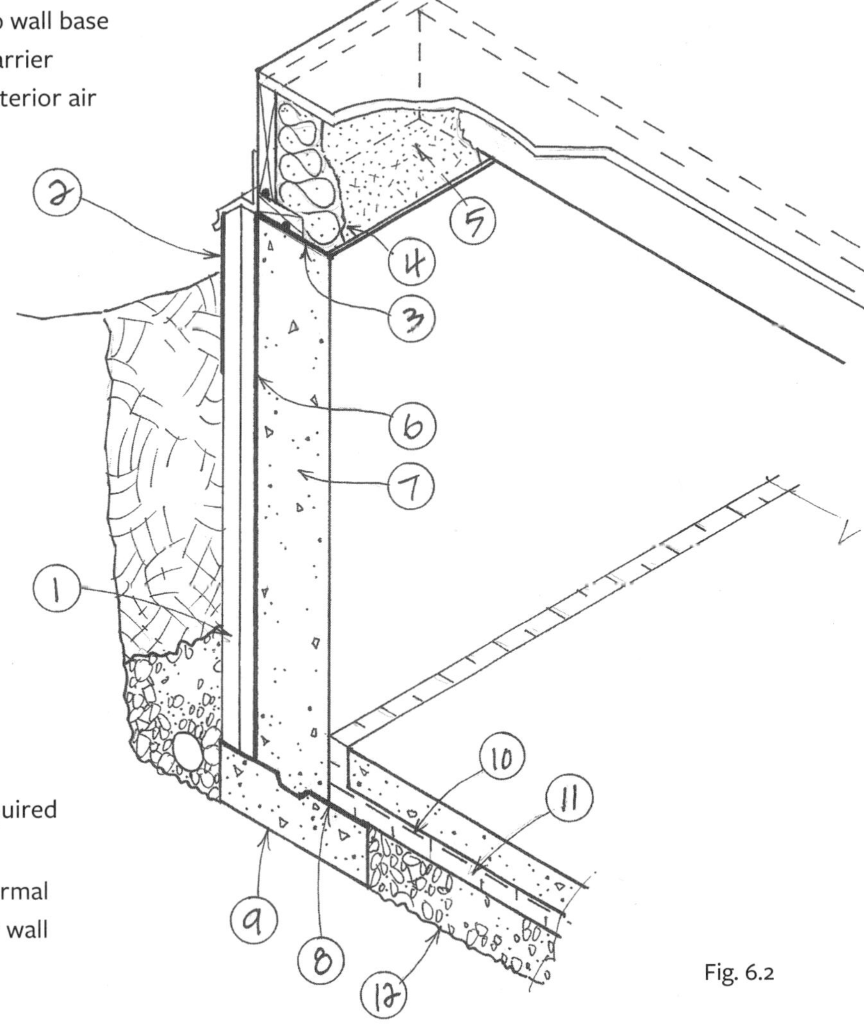

Fig. 6.2

Basement with Interior Insulation

When a basement is fitted with interior insulation, the water control layer may be placed to the inside of a foundation wall, with the concrete wall exposed directly to the soil; in these cases, a subgrade liquid-applied "dampproofing" treatment (often an asphalt-based coating) is applied to reduce the amount of moisture driven into the concrete from the ground. Note that dampproofing is not designed to be a water barrier and will not resist hydrostatic water pressure! A full water control layer located to the exterior may be desired as a second layer of water control in buildings with significant groundwater pressure.

Figure 6.3 shows some details for designing a basement with interior insulation:

Advantages

- Insulation is well-protected from exterior damage (i.e., UV, fire, insects).
- Foundation can be brought above grade to help protect vulnerable bottom-of-wall details from surface water, splashback, etc.
- Relatively easy to create thermal bridge-free construction and tie into both band-joist and sub-slab insulation.
- Standard design uses common materials and trades practices.
- Basements may provide economical and/or practical space in the building design, whether finished or used as storage, mechanical, or utility space.

Disadvantages

- Interior insulation results in cooler foundation walls, increasing risk of condensation and requiring a robust interior air barrier, basement humidity control, and moisture-tolerant construction in the event of incidental wetting. This approach is a more moisture-vulnerable approach than exterior-insulated foundation walls.
- Wet-service location requires more ecologically damaging materials (i.e., foam, mineral board), which may conflict with project values; bio-based materials may be higher-risk in the event of flooding or moisture damage due to poor design detailing.
- Cost may be higher if additional finishing and protection of insulation is required beyond what is needed for the intended use of the space (i.e., storage, utility).
- Thermal mass benefits of masonry wall are lost.

1. Capillary break (i.e., structural gasket, liquid-applied sealant) to ensure liquid moisture in the foundation wall cannot wick up into sensitive parts of the wall assembly. Gaskets are a great solution, as they can also serve to control air through this junction as part of the air control layer.
2. Sealant to provide continuity of air control layer: a) from foundation wall to mud sill (if not provided by capillary break), b) from mud sill plate to band/rim joist, c) from band/rim joist to subfloor, d) from subfloor to wall base plate, e) from wall base plate to interior air barrier materials, and/or f) from wall base plate to exterior air barrier materials.
3. Band-joist insulation (i.e., cellulose, batt, spray foam).
4. Water-control layer placed directly against the foundation wall. Materials for this layer include peel-and-stick membranes or liquid-applied moisture barriers such as asphaltic coatings (dampproofing) or other masonry sealants.
5. The concrete slab and walls serve as the air barrier, providing that all penetrations and transitions, such as plumbing and electrical stubs, sump pump housings, and doors and windows, are installed to be airtight.
6. Rigid vapor- and air-impermeable foam board insulation (i.e., foil-faced polyiso).
7. Wood-frame wall (use rot-resistant wood such as cedar or locust), insulated with vapor-permeable cavity insulation.
8. Paperless or moisture resistant (MR) gypsum board, cement board, or other moisture-tolerant material with vapor semi-permeable finish (i.e., latex paint) to allow inward drying potential and vapor control.
9. Capillary break — liquid-applied or sheet membrane — between wall and footing is required when footing is in direct contact with soil.
10. Concrete footing below frost depth.
11. Water/vapor control layer below the slab is a sub-slab-grade polyethylene membrane, and may act as a radon barrier. All seams must be sealed, and membrane must be permanently sealed to foundation walls.
12. Sub-slab rigid insulation (i.e., high-density mineral board, high-density EPS or XPS foam). Slab insulation levels must be higher for slabs fitted with radiant heating systems. Note continuity with wall insulation to avoid thermal bridging through footing. Termite control may be a concern for subgrade insulation in certain regions.
13. Granular capillary break/drainage pad (i.e., crushed stone or gravel) reduces bulk moisture wicking into slab and facilitates interior drainage and soil gas mitigation systems, when present.

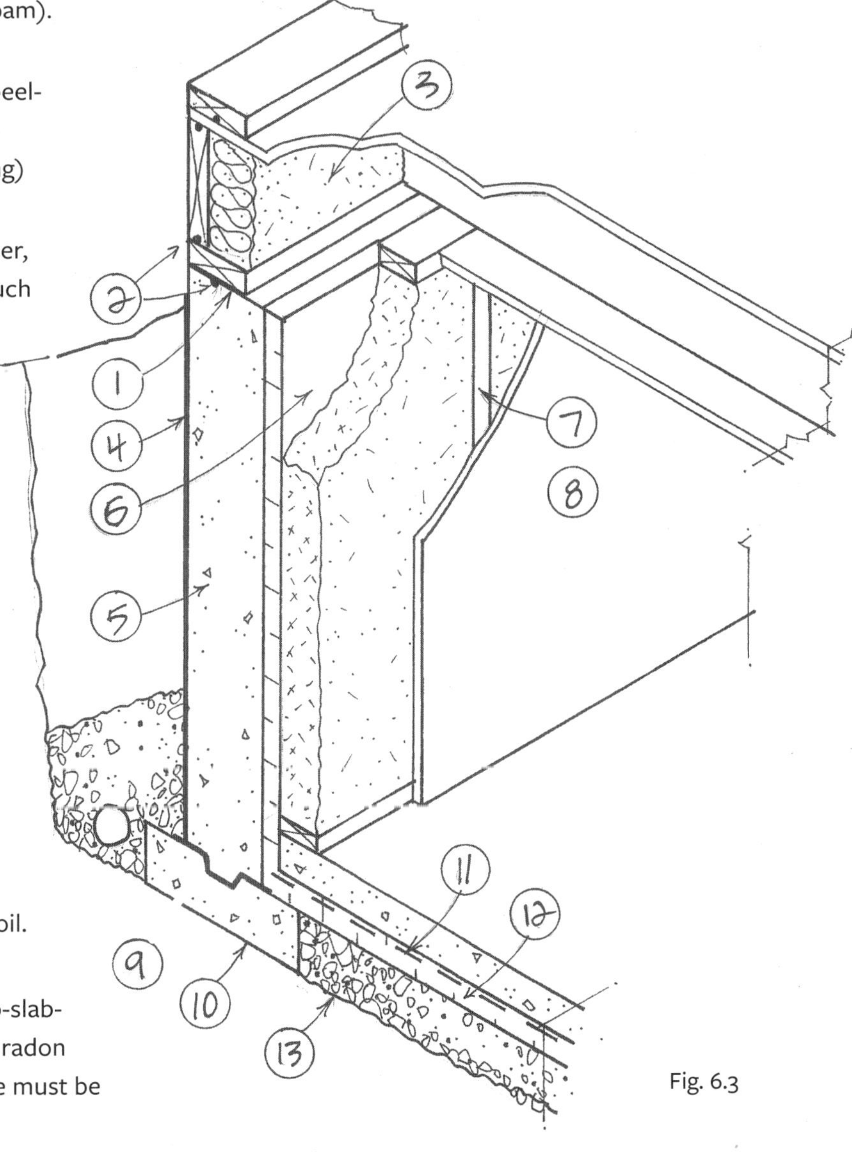

Fig. 6.3

Crawl Spaces

Crawl spaces are short, unfinished basements, and should be treated as such; although unfinished, they should be unvented and kept within the building enclosure. Vented crawl spaces, while code-approved (and even code-mandated in some cases), create significant moisture problems by allowing warm humid air to enter during the summer and condense against cold surfaces; in the winter they allow cold air to enter, further cooling surfaces without providing significant drying potential. Therefore, unvented crawlspaces are recommended, with conditioning to be provided by the building's HVAC system when possible, or supplemental dehumidification as needed.[1] In some cases, such as retrofits, it becomes necessary to isolate the crawl space from the building enclosure, in which case vapor, air, and thermal control layers should be detailed accordingly into the floor assembly (see Figure 6.6 Pier Foundation, below).

Advantages

- Less expensive to construct (less excavation, don't require finishing).
- Flexible design can work with a variety of different flooring systems.
- No direct ground contact with thermal enclosure allows for more flexible material choice (if foundation walls are not part of thermal enclosure).
- Standard construction practice uses commonly found materials and trade support, generally without custom design or engineering requirements.

Disadvantages

- Unfinished nature often results in an inadequate job isolating the building interior from the crawl space, resulting in moisture and thermal penalties.
- Uncontrolled humidity release from saturated soils, surface water flooding, and groundwater intrusion can result in mold, high building humidity, and framing rot.
- Many codes and much common practice dictates the "venting" of crawl spaces. In practice, this serves to introduce very cold air in the winter, causing thermal losses and discomfort, and very warm humid air in the summer that, upon hitting the relatively cool surfaces of the ground and building framing, condenses and introduces far more moisture than is released.
- Even when fully sealed and vapor-controlled, the crawl space may be difficult to keep at an appropriate humidity level without mechanical support (i.e., dehumidification, forced ventilation).

1. Rigid insulation suitable for direct contact with soil (i.e., mineral board, fiberglass board, high-density EPS or XPS foam) — note that this layer acts as a water control layer (and vapor control, if the material itself is vapor-impermeable) as well as thermal control. Termite control may be a concern for subgrade insulation in certain regions.
2. Exterior treatments to protect above-grade insulation from insects, UV degradation, and erosion, installed beneath flashing between above-grade wall and foundation.
3. Capillary break (i.e., structural gasket, liquid-applied sealant) to ensure liquid moisture in the foundation wall cannot wick up into sensitive parts of the wall assembly. Gaskets are a great solution, as they can also serve to control air through this junction as part of the air control layer.
4. Sealant to provide continuity of air control layer: a) from foundation wall to mud sill (if not provided by capillary break), b) from mud sill plate to band/rim joist, c) from band/rim joist to subfloor, d) from subfloor to wall base plate, e) from wall base plate to interior air barrier materials, and/or f) from wall base plate to exterior air barrier materials.
5. Exterior wall rigid insulation continued over floor assembly (see "Wall Assemblies," below for insulation strategy).
6. Band-joist insulation (i.e., cellulose, batt, spray foam).
7. Water control layer placed directly against the foundation wall. This redundancy backs up the water control role of the insulation and provides vapor control in the instance of a vapor-permeable insulation such as mineral board. Materials for this layer include peel-and-stick membranes or liquid-applied moisture barriers.
8. Concrete walls serve as the air barrier, providing that all penetrations and transitions, such as plumbing and electrical stubs, sump pump housings, and doors and windows, are installed to be airtight.
9. Capillary break — liquid-applied or sheet membrane — between wall and footing is required when footing is in direct contact with soil.
10. Concrete footing below frost depth.
11. Water/vapor control layer over the ground may be specified to act as a radon barriers as well. Note that a suitably rugged membrane should be used to withstand damage during construction, or if future access is expected. All seams must be sealed, and membrane must be permanently sealed to foundation walls.

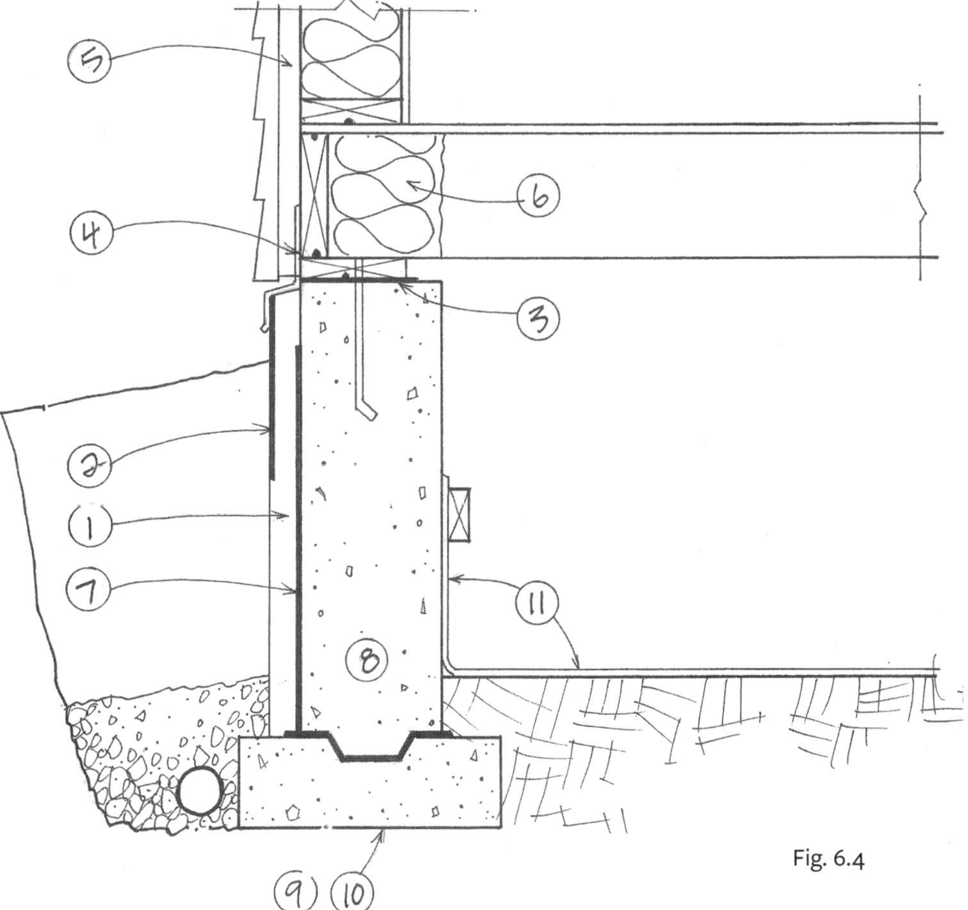

Fig. 6.4

Frost-protected Shallow or Monolithic Slab Foundation

Frost-protected shallow foundations (FPSF) rely on vertical perimeter and horizontal wing insulation to prevent frost-heaving; the insulation keeps the soil below the footings from freezing, and it requires good drainage at the footing plane to reduce free water from potentially infiltrating the foundation.[2] Monolithic slabs are one type of FPSF; in these, the footing is integrally formed into the slab, known as a *thickened-edge slab.* When the entire slab is poured at a uniform structural thickness, it is known as a *raft slab foundation.*

1. Rigid insulation suitable for direct contact with soil (i.e., mineral board, fiberglass board, high-density EPS or XPS foam). Note that this layer acts as an additional water control layer (and vapor control, if the material itself is vapor-impermeable) as well as thermal control. Wing insulation as needed to ensure above-freezing conditions below slab footing; vertical insulation is used to prevent thermal bridging through the slab edge. Termite control may be a concern for subgrade insulation in certain regions.
2. Exterior treatments to protect above-grade insulation from insects, UV degradation, and erosion, installed beneath flashing, between above-grade wall and foundation.
3. Capillary break (i.e., structural gasket, liquid-applied sealant) to ensure liquid moisture in the foundation cannot wick up into sensitive parts of the wall assembly. Gaskets are a great solution, as they can also serve to control air through this junction as part of the air control layer.
4. Sealant to provide continuity of air control layer: a) from slab to wall base plate (if not provided by capillary break), b) from wall base plate to interior air barrier materials, and/or c) from wall base plate to exterior air barrier materials.
5. The concrete slab serves as the air barrier, providing that all penetrations and transitions, such as plumbing and electrical stubs, and sump pump housings are installed to be airtight.
6. Water/vapor control layer below the slab and extending up the slab edge to tie into slab-to-wall transition. This layer is a sub-slab-grade polyethylene membrane, and may act as a radon barrier. All seams must be sealed, and membrane must be permanently sealed to foundation walls.
7. Sub-slab rigid insulation (i.e., high-density mineral board, high-density EPS or XPS foam). Slab insulation levels must be higher for slabs fitted with hydronic heating systems. Termite control may be a concern for subgrade insulation in certain regions.
8. Structural insulation (25 psi minimum, i.e., XPS, high-density EPS) beneath footings.
9. Granular capillary break/drainage pad (i.e., crushed stone or gravel) reduces bulk moisture wicking into slab and facilitates interior drainage and soil gas mitigation systems, if needed.

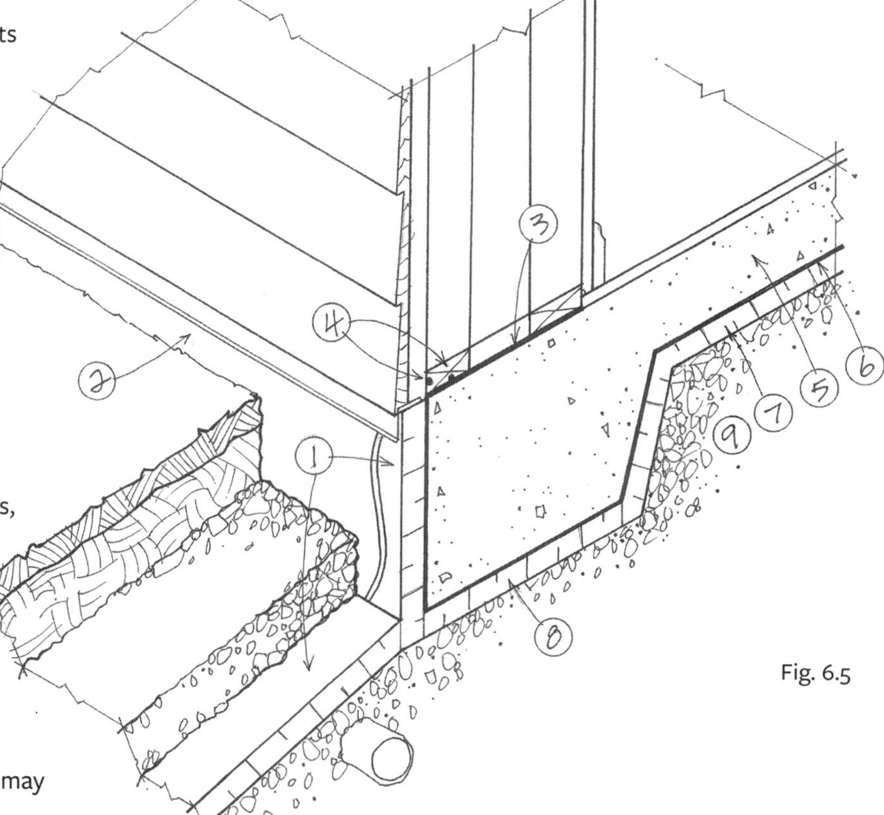

Fig. 6.5

Advantages

- Slabs-on-grade are often less expensive, requiring less excavation, formwork, and materials.
- Slabs may be easily and inexpensively finished, and they can receive a variety of different flooring materials.
- Continuous layers are straightforward to execute as there are fewer transitions than with other foundation styles.
- Ecological damage may be reduced by lessening the need for excavation and using less poured concrete, both of which are significant contributors to embodied carbon.
- Details are reliable and have been time-tested for even the coldest climates.

Disadvantages

- Services such as plumbing and electrical must be designed and installed prior to foundation pour, and remodeling becomes significantly more complicated, as services are located below the slab.
- Wet-service location requires more ecologically damaging materials (i.e., foam, mineral board), which may conflict with project values.
- Protecting bottom-of-wall adequately may require masonry stub walls or other details to provide sufficient distance between the ground and the vulnerable parts of the assembly.
- Many trades don't understand how to detail slabs-on-grade correctly, or don't trust they can be used reliably in freeze-prone locations.

Insulated Concrete Forms and Sandwich Walls

A foundation wall system that is gaining popularity is the insulated concrete form, or ICF. In this system, foundation walls are built using small, modular, permanent forms that have insulation incorporated into the them, with appropriate reinforcement bar placed as the forms go up. The cores of these forms are then poured with a standard concrete ready-mix. They are also used for above-grade walls.

The control layers are similar to those in an exterior basement wall — with a few differences: the water and vapor control layer is placed outboard of the exterior insulation, which is now formed into the wall. The massive wall is inherently airtight. The thermal control layer is split between the interior and exterior of the wall, as the form is insulated on both sides. This means the concrete is separated from the interior, and does not perform as thermal mass. The interior surface is warm, however, and there is very low condensation potential with this system.

Advantages include excellent durability, moderate-to-high levels of insulation, simple-to-build forms that do not require specialty equipment, and a wide range of styles and materials. Disadvantages include higher cost, no thermal mass benefit, and very few foam-free options. Exposed insulation must still be protected above grade (and below grade in termite-prone regions) and on the interior. Insulation levels may not be adequate in colder climates for some systems.

Another approach for both foundation and above grade walls is the sandwich, or CIC (for concrete-insulation-concrete) wall. Built like a standard poured wall, a layer of insulation is placed into the center of the form, with factory-engineered structural ties connecting the two outer *wythes* of concrete together to operate as a whole structural system.

The control layers are the same as those in an exterior-insulated basement wall, except that the insulation is set in the middle of the wall, and not against the exterior.

Advantages include enhanced durability because the insulation is at no risk of damage. Standard forms can be used, and training is minimal for inexperienced contractors. The mass benefit of the inner wythe of concrete is preserved, and only minimal finishing is required. Disadvantages are high costs with few options of vendors, few foam-free options, and limitations to insulation levels, which may make them inadequate for cold climates.

Piers

Pier foundations fully elevate the enclosure off of the ground, formally decoupling the building from the earth. In some ways, this simplifies things by reducing ground and surface moisture exposure. In other ways, things are made more complicated by having the bottom layer of an enclosure exposed to air rather than buffered by more moderate ground temperatures; this is especially true in more extreme hot and cold climates. The floor framing becomes the bottom horizontal plane of the building enclosure, and all moisture and thermal control layers must be integrated into this assembly. The pier foundation system then serves a purely structural role. To that end, one can think of the floor as a wall tipped horizontally, controlling heat, air, and moisture across the same differences in temperature and humidity between the interior and exterior of the building. Air-sealing the entirety of this floor assembly is critically important to reduce condensation potential as well as unwanted heat transfer. Vapor control strategies can vary; these are discussed further in the section that follows about walls.

 ***Advantages***

- Perhaps the least expensive of all foundation types.
- Perhaps the lowest ecological damage, by virtue of reduced material use and excavation.
- Foundation can be brought above grade to help protect vulnerable bottom-of-wall details from surface water, splashback, etc. Buildings can even be placed in flood-prone locations, allowing water to flow below the enclosure.
- Installation is relatively simple and does not require complicated formwork, carefully installed layers, or sub-slab electrical or plumbing services.

 ***Disadvantages***

- All water, air, vapor, and thermal control must be integrated into the floor assembly.
- Most common detailing involves use of exterior rigid foam board or spray foam cavity insulation to reduce condensation within the assembly. Foam-free solutions require more careful detailing and carry a higher risk, especially in cold and humid climates.
- Additional thermal stress on the building due to decoupling from temperature-buffered earth.
- Pier design and installation is custom and specific to structural loads of the building and topography of the site; careful detailing, such as drainage and/or horizontal in-ground insulation, is required to avoid frost heaving, common in cold climates.
- Bringing water up from below grade in freezing climates presents a significant challenge in keeping water lines from freezing between the ground and the building.

1. Pier foundation — poured concrete, masonry block units, treated wood, metal piling, helical screw, or other system of choice as required to match building structural loads, soil type, and hydrology.
2. Footing below frost depth, as required by foundation system.
3. Capillary break (i.e., structural gasket, liquid-applied sealant) between the ground-coupled pier (if concrete) and the wood beam to ensure liquid moisture in the pier cannot wick up into sensitive parts of the structure above.
4. Sealant to provide continuity of air control layer: a) from foundation pier to band/rim joist, b) from band/rim joist to subfloor, c) from subfloor to wall base plate, d) from wall base plate to interior air barrier materials, and/or e) from wall base plate to exterior air barrier materials.
5. Exterior wall rigid insulation continued over rim joist (see "Wall Assemblies," below, for insulation strategy).
6. Rigid insulation below floor frame (to reduce thermal bridging) radiantly decouples the wood framing from the ground, and boosts assembly R-value. Vapor- and air-tight foam board is often recommended to reduce vapor entry during the summer or in hot-humid climates; if an air-permeable insulation is used (i.e., mineral board), an airtight membrane must be installed, such as sealed plywood or a plastic air barrier sheet. This is particularly important for vapor- and air-permeable cavity insulation.
7. Vapor- and air-permeable cavity insulation should fully fill the cavity, making contact with the underside of the subfloor. Alternately, closed-cell spray foam can be used to seal the underside of the subfloor, providing air, vapor, and thermal control.

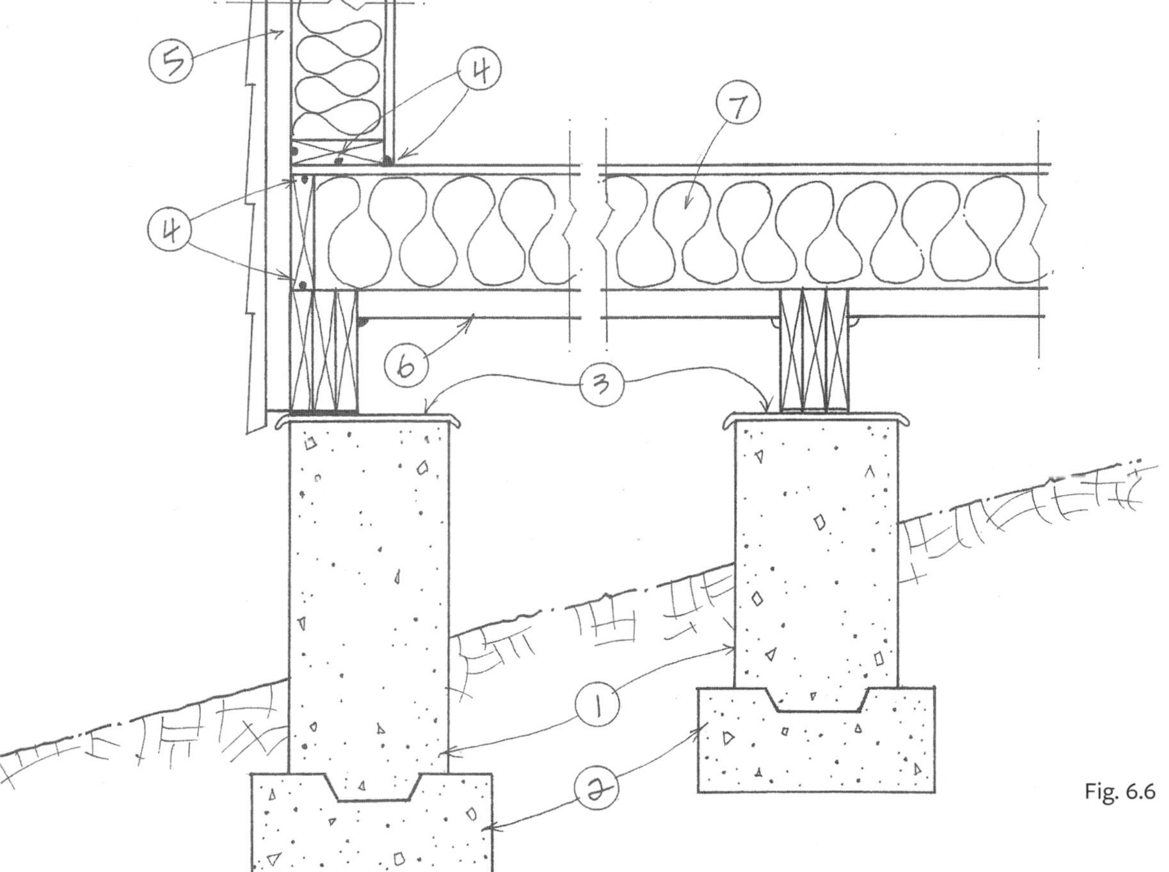

Fig. 6.6

Wall Assemblies

For the purposes of this book, we are restricting our conversation to framed walls, which are the most common style in residential construction in North America. Even within this range, there are myriad approaches. We have selected a handful that cover moderate and high levels of thermal performance featuring different strategies for vapor control and material selection.

Note that many other options for creating high-performance wall assemblies exist, including insulated concrete forms, SIPs (including prefabricated foam-free panels), different types of mass walls, masonry and other cladding types, and various innovative framing approaches that optimize thermal performance and durability. The approaches presented here can be used to inform strategies for other wall types.

Insulated Stud Wall with Exterior Vapor-Permeable Insulation

In the simplest approach to vapor control, we look for the direction of the dominant vapor drive, and place the vapor and air control layers on the humid side. This logic is pretty straightforward: we want to control the amount of vapor entering the assembly by locating the control layers where diffusion into the building assembly is occurring, and we want to allow any humidity to diffuse out (dry) in the direction that it is driven. In the majority of cases, this results in an insulated stud wall with exterior vapor-permeable continuous insulation; an optional vapor control layer is climate dependent.

Advantages

- Most standard style of residential construction in North America; inexpensive to build.
- Flexible options for insulation allows for range of insulation and finish materials, including the use of low-impact materials (i.e., cellulose, fiberboard).
- Open-vapor profile allows for easy adoption in a variety of climates; vapor control can easily be introduced as required for cold or hot-humid climates.
- Rainscreen water control system can be easily integrated for most claddings.

Disadvantages

- Common practice doesn't account for good air barrier detailing; extra attention must be paid here.
- Attention must be paid to vapor profile of materials relative to climate — for example, do not use a vapor-closed interior wall covering, such as vinyl wallpaper, in a hot-humid climate!
- May not provide sufficient insulation for cold climates without significant increase of wall thickness, which may complicate trim and siding detailing, and may not provide sufficient external R-value to keep sheathing above dew point.

From outside to inside:

1. Cladding.
2. Gap for drainage and drying — recommended in all cases, and required in cases where exterior drying must be optimized or a capillary break behind the cladding is necessary.
3. Vapor-semi-permeable or permeable insulated sheathing, such as mineral board, unfaced fiberglass board, or fiberboard (optional). This will optimize the thermal performance of the wall and help keep the sheathing above the dew point, reducing risk of condensation damage in cold climates. Airtight materials may be sealed to function as an air control layer.
4. WRB, or "housewrap," such as 15# asphalt-impregnated felt or plastic membranes. If this layer is also designed to act as an air barrier, a membrane should be selected and installed that can serve as both a water and air control layer. Liquid-applied membranes are starting to appear in the residential market. In cold and mixed climates, these membranes are vapor-permeable, allowing drying to the exterior; in hot-humid climates vapor-semi-impermeable or impermeable membranes are used, where the vapor drive is predominantly toward the interior.
5. Flashing at the various transitions within the wall plane (i.e., windows, vent ports, exterior outlets and lights, porch framing) and at the foundation at the bottom of the wall.
6. Vapor semi-permeable structural sheathing, such as OSB, plywood, or diagonal board sheathing. Plywood and OSB can be sealed to perform as an air barrier.
7. Wood frame, 2x4/6/8, insulated with vapor-permeable cavity insulation.
8. Airtight vapor-variable or vapor-semi-impermeable membrane, in cold climates. Batt facing may serve as a vapor control layer in cold climates, but not as an air barrier. In mixed and hot-humid climates, there should be no vapor control here.
9. Gypsum wallboard (GWB), which can be detailed as an air barrier as an optional replacement for the membrane above, finished with a semi-permeable latex paint (or use a semi-impermeable paint to replace the vapor control membrane in cold climates).

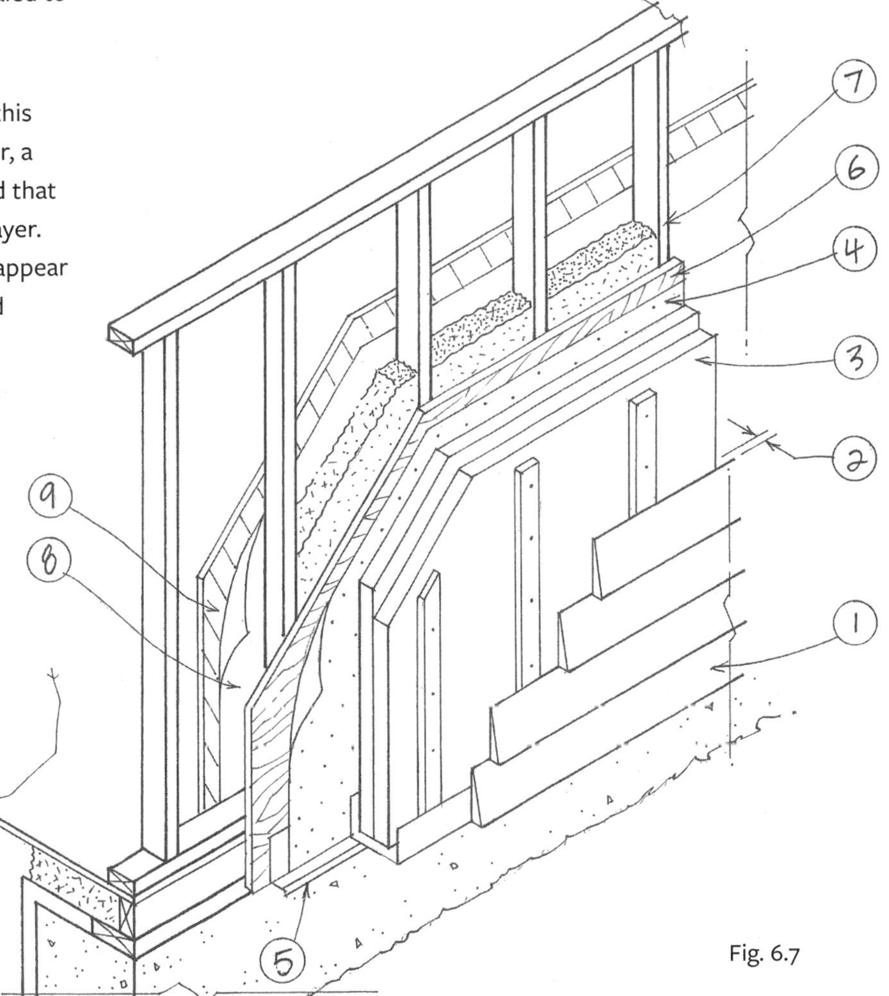

Fig. 6.7

Insulated Stud Wall with Exterior Vapor-closed Continuous Insulation

Dubbed the "perfect wall" by building scientist Joe Lstiburek,[3] all of the control layers — including a vapor control layer — are oriented to the exterior of the structure. How is this possible in a cold climate without allowing outward-driven vapor to condense on this "cold-side" vapor barrier? The trick is that enough of the insulation is continuous outboard of this vapor barrier, relative to the in-cavity R-value, keeping it above the dew point, and allowing drying to the interior ("enough" varies based on your region, with some guidelines published in the IRC[4]). The vapor control layer is somewhere in the middle of the assembly during heating seasons, thermally speaking. The balance of insulation is very important here — if you have significantly more insulation in the frame cavity (inboard of the vapor control layer) than outboard, the vapor control layer runs the risk of dropping below the dew point. Many designers further advocate for the use of high-density closed-cell spray foam or XPS foam board as the insulation, as it will then function as both water control, vapor control, thermal control, and air control, all in one material, if detailed correctly.[5] From a building science perspective, this illustrates the popularity of foam in high-performance construction, providing that larger-context ecological, toxicological, and climate-related impacts are ignored or discounted. Regardless of the insulation type, if applied correctly, this is a very durable strategy, as the structural elements of the assembly are kept entirely within the control layers — warm and dry.

 ***Advantages***

- Structure is kept warm and dry; can be applied in any climate without altering details.
- Wall cavity can perform as a service cavity, allowing for future service work (i.e., electrical) without risk of damaging the air barrier.
- Foam insulation can provide the work of all four layers — water control, vapor control, air control, and thermal control — if detailed correctly.
- Additional advantages of this structural approach (ubiquity, cost, flexibility), as described in previous example.

 ***Disadvantages***

- Reliance on foam has significant ecological, health, and climate impacts.
- Detailing must be meticulous to ensure foam can remain air- and water-tight over time; redundant layers may be required.
- In cold climates, balance of insulation between the cavity insulation and the continuous foam insulation must be struck to ensure the wall sheathing stays above the dew point.
- May not provide sufficient insulation for cold climates without dramatic increase of wall thickness.

From outside to inside:

1. Cladding.
2. Gap for drainage and drying — recommended in all cases, and required in cases where exterior drying must be optimized or a capillary break behind the cladding is necessary.
3. Rigid board insulation, such as EPS, XPS, or polyisocyanurate foam, mineral board, fiberglass, or cork. Airtight board insulation may also serve as an air barrier, and vapor-semi-impermeable board insulation may triple its function as the vapor control layer. Hydrophobic board insulation could theoretically also act as the fourth layer of water control, although common practice in wet climates is to use a WRB membrane to facilitate more reliable flashing.
4. WRB, or "housewrap," such as 15# asphalt-impregnated felt or plastic membranes. If not already provided by the board insulation, this membrane can be specified and detailed to be both an air and vapor control layer.
5. Flashing at the various transitions within the wall plane (i.e., windows, vent ports, exterior outlets and lights, porch framing) and at the foundation at the bottom of the wall.
6. Sheathing, such as OSB, plywood, or diagonal board sheathing. Plywood and OSB can be sealed to perform as an air barrier.
7. Wood frame, 2×4 / 6/8, insulated with vapor-permeable cavity insulation of choice (i.e., dense-pack cellulose, unfaced batts). Remember that the R-value of this layer must be kept in balance with the rigid board insulation, according to climate.
8. Gypsum wallboard (GWB), which can be detailed as an air barrier, finished with a semi-permeable or permeable paint (i.e., clay, lime, latex).

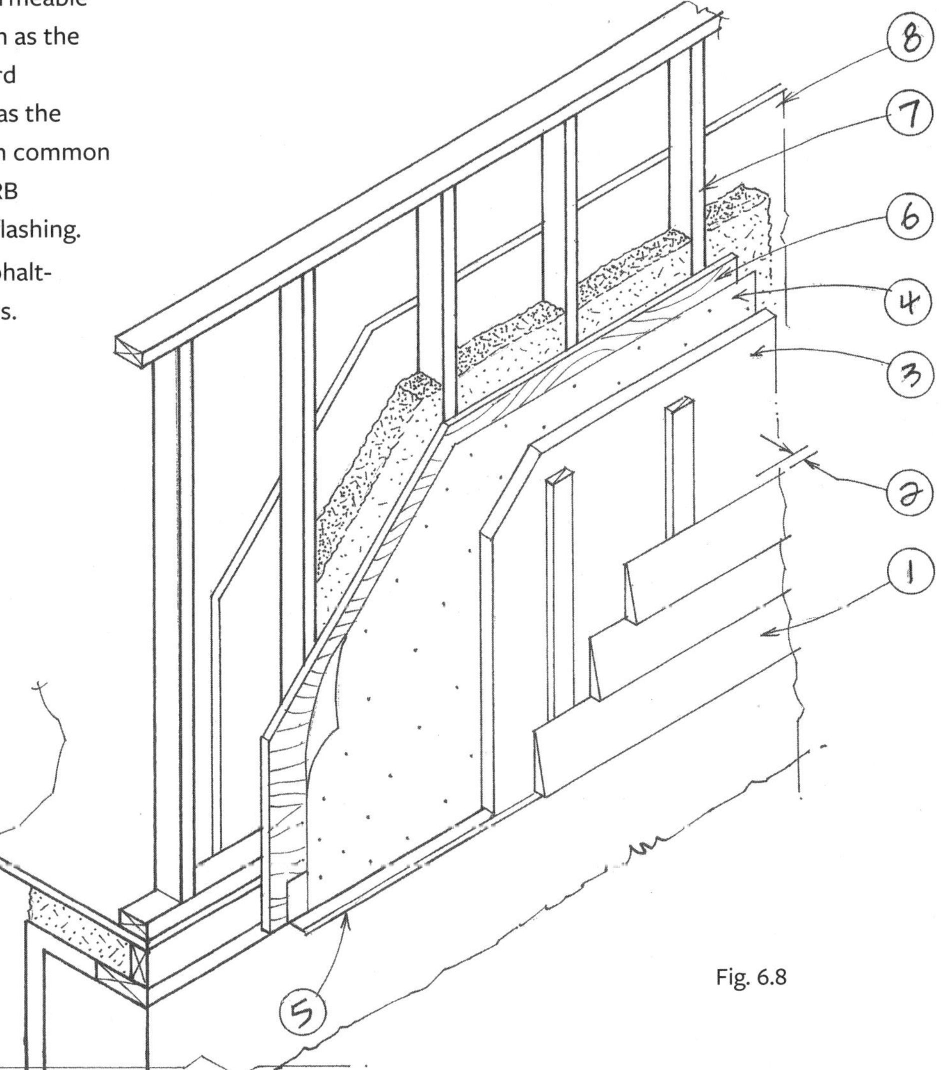

Fig. 6.8

High-performance Vapor Flow-through Wall

High R-value wall assemblies have a greater risk for moisture damage in cold climates, as layers outboard of the insulation will be colder longer, increasing potential for condensation and reducing any drying potential. One approach for vapor management in a high-performance super-insulated wall assembly is to allow vapor to diffuse all the way through the assembly in either direction. This is the approach most commonly used in the world of natural building, but it bears just as much relevance in assemblies built with more conventional vapor-permeable materials. There is a caveat here, however — in very cold regions (IECC Climate Zones 6 or colder), if there is not sufficient continuous outboard insulation, a vapor "throttle," in the form of some vapor control — like a vapor-variable membrane or surface coating — is required toward the middle or interior of the assembly, as in these extreme cold conditions even relatively lower levels of diffusional-driven vapor can cause damage in uninsulated exterior sheathing.[6] Further, a rainscreen is recommended for these assemblies to ensure sufficient outward drying potential is available (depending on the leakiness of the cladding), particularly in cold climates. Like all super-insulated approaches, a high-quality air barrier is required to avoid condensation problems.

 ***Advantages***

- Flexible approach to vapor allows for maximum drying, can be applied in all but the coldest climates without alteration.
- Allows for the use of the most ecologically benign materials in a moisture-durable assembly.
- Superior thermal performance yields greater comfort, reduced ecological impact, lower operating costs.
- Incorporates the "5 D's" (*Design, Deflection, Drainage, Deposit, and Drying*) effectively, relying on a balanced approach to moisture management.

 ***Disadvantages***

- In coldest climates, vapor drive may overpower drying potential during winter, so a vapor throttle may be required. Incidental condensation may occur, particularly in cold climates, and appropriate materials must be selected to provide safe storage (i.e., board or plywood sheathing, clay or lime plaster).
- Attention must be paid to vapor profile of materials relative to climate — for example, do not use OSB sheathing in a cold climate, as OSB does not allow sufficient drying and cannot adequately handle incidental wetting.
- Super-insulated walls require a robust air barrier to reduce condensation risk.

From outside to inside:

1. Cladding.
2. Rainscreen: ¼″–1″ air space for drainage and drying, screened and flashed at bottom of wall (ideally, the top of the wall is also vented) — optimizes exterior drying and provides a capillary break behind the cladding.
3. (not shown) WRB, or "housewrap," such as 15# asphalt-impregnated felt or plastic membranes. If this layer is also designed to act as an air barrier, a membrane should be selected and installed that can serve as both a water and an air control layer. While more common in commercial applications, liquid-applied membranes are starting to appear in the residential market as well.
4. Flashing at the various transitions within the wall plane (i.e., windows, vent ports, exterior outlets and lights, porch framing) and at the foundation at the bottom of the wall.
5. (not shown) Vapor semi-permeable structural sheathing, such as OSB, plywood, or diagonal board sheathing. Plywood and OSB can be sealed to perform as an air barrier. Most OSB does not support sufficient drying and is particularly susceptible to moisture damage; select plywood or board sheathing in cold climates.
6. Alternative: replace items 1 to 5 with an exterior permeable plaster system (i.e., 2- or 3-coat lime or clay plaster), sealed with a vapor-permeable water control coating (i.e., limewash, siloxane), or replace 3 to 5 for rough coat plaster with rainscreen and cladding, for natural building assemblies.
7. Outer 2×4 structural wood frame (use 2×3 if non-structural).
8. Insulated with vapor-permeable cavity insulation of choice (i.e., dense-pack cellulose, hempcrete, straw-clay).
9. Inner 2×3 non-structural wood frame (use 2×4 if structural).
10. (not shown) Alternative: replace items 7, 8 and/or 9 with straw bales for straw wall construction.
11. Gypsum wallboard (GWB) or interior clay, gypsum, or lime plaster system, detailed as an air control layer, finished with a semi-permeable or permeable paint (i.e., latex, clay, lime). Note: in cold or very cold climates, use a semi-impermeable paint to provided needed vapor control.

Fig. 6.9

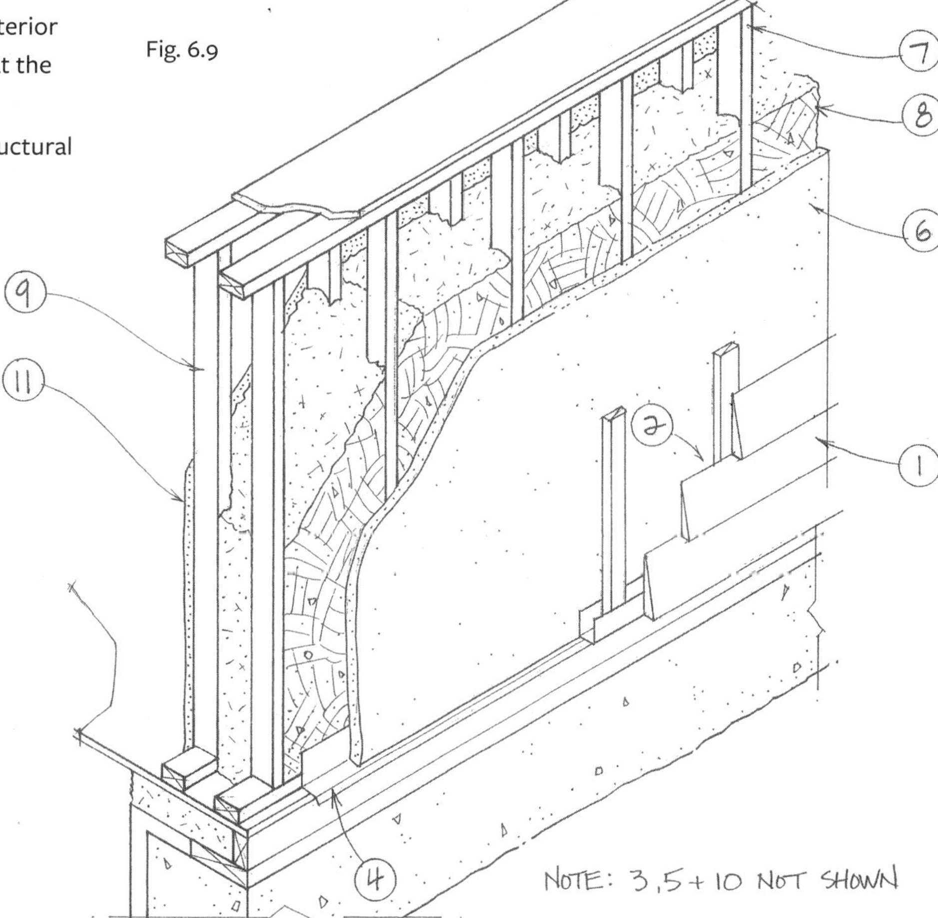

High-performance Double-stud Wall with Vapor Throttle

Another strategy, often found in high-performance assemblies, is to place a vapor control layer somewhere in the middle of the assembly, with vapor-permeable layers to the inside and outside of the assembly. This "throttle" helps limit vapor flowing all the way through the assembly, but will still allow for drying in either direction as conditions require.[7] An example of this would be plywood (or OSB sheathing or a vapor control membrane) installed to the outer face of the inner studs of a double-stud wall. Like all super-insulated approaches, a high-quality air barrier is required to avoid condensation problems; the vapor control layer often doubles as the air control layer.

 ***Advantages***

- Assembly can dry in either direction, allowing for design flexibility and application in all climates.
- Supports the durable and safe use of ecologically benign materials in even very cold regions.
- Superior thermal performance yields greater comfort, smaller ecological impact, and lower operating costs.
- Wall cavity can perform as a service cavity, allowing for future service work (i.e., electrical) without risk of damaging the air barrier.
- Incorporates the "5 D's" effectively, relying on a balanced approach to moisture management.

 ***Disadvantages***

- Attention must be paid to vapor profile of materials relative to climate — for example, do not use OSB exterior wall sheathing in a cold climate, as OSB does not allow sufficient drying and cannot adequately handle incidental wetting. (Note that use of OSB in this scenario is not used as exterior wall sheathing, but as an interstitial air and vapor control layer).
- Depending on the wall assembly, incorporation of an interstitial vapor throttle may not be trivial, and it can add expense and complication to the construction process. This approach may not be compatible with a straw bale wall assembly.
- Super-insulated walls require a robust air barrier to reduce condensation risk.

A note on yet another approach, which is the *double vapor barrier* assembly: these assemblies feature vapor barriers on both sides of the assembly, and offer little to no drying potential. These types of assemblies, while not entirely uncommon, are to be avoided because they create a higher risk. Common examples are flat roofs featuring interior vapor barriers and walls with continuous outboard XPS insulation and an interior vapor barrier. In some cases they can work, but they are not recommended.

From outside to inside:

1. Cladding.
2. Rainscreen: ¼″–1″ air space for drainage and drying, screened and flashed at bottom of wall (ideally, the top of the wall is also vented) — optimizes exterior drying and provides a capillary break behind the cladding.
3. WRB, or "housewrap," such as 15# asphalt-impregnated felt or plastic membranes. If this layer is also designed to act as an air barrier, a membrane should be selected and installed that can serve as both a water and an air control layer. While more common in commercial applications, liquid-applied membranes are starting to appear in the residential market as well.
4. Flashing at the various transitions within the wall plane (i.e., windows, vent ports, exterior outlets and lights, porch framing) and at the foundation at the bottom of the wall.
5. Vapor semi-permeable structural sheathing, such as OSB, plywood, or diagonal board sheathing. Plywood and OSB can be sealed to perform as an air barrier. Most OSB does not support sufficient drying and is particularly susceptible to moisture damage; select plywood or board sheathing in cold climates.
6. (Not shown) Alternative: replace items 1 to 5 with an exterior permeable plaster system (i.e., 2- or 3-coat lime or clay plaster), sealed with a vapor-permeable water control coating (i.e., limewash, siloxane), for natural building assemblies.
7. Outer 2×4 structural wood frame (use 2×3 if non-structural).
8. Vapor-permeable cavity insulation of choice (i.e., dense-pack cellulose, hempcrete, straw-clay).
9. Vapor and air control layer, such as sealed plywood, OSB, or an airtight vapor-semi-impermeable or vapor-variable membrane.
10. Inner 2×3 non-structural wood frame (use 2×4 if structural). Insulated with vapor-permeable cavity insulation of choice (i.e., dense-pack cellulose, hempcrete, straw-clay). Note that this cavity, being located inboard of the air control layer, may now act as a service cavity, allowing installation and modification of electric and other services without risk of penetrating the air barrier.
11. Gypsum wallboard (GWB) or interior clay, gypsum, or lime plaster system, finished with a semi-permeable or permeable paint (i.e., latex, clay, lime).

Fig. 6.10

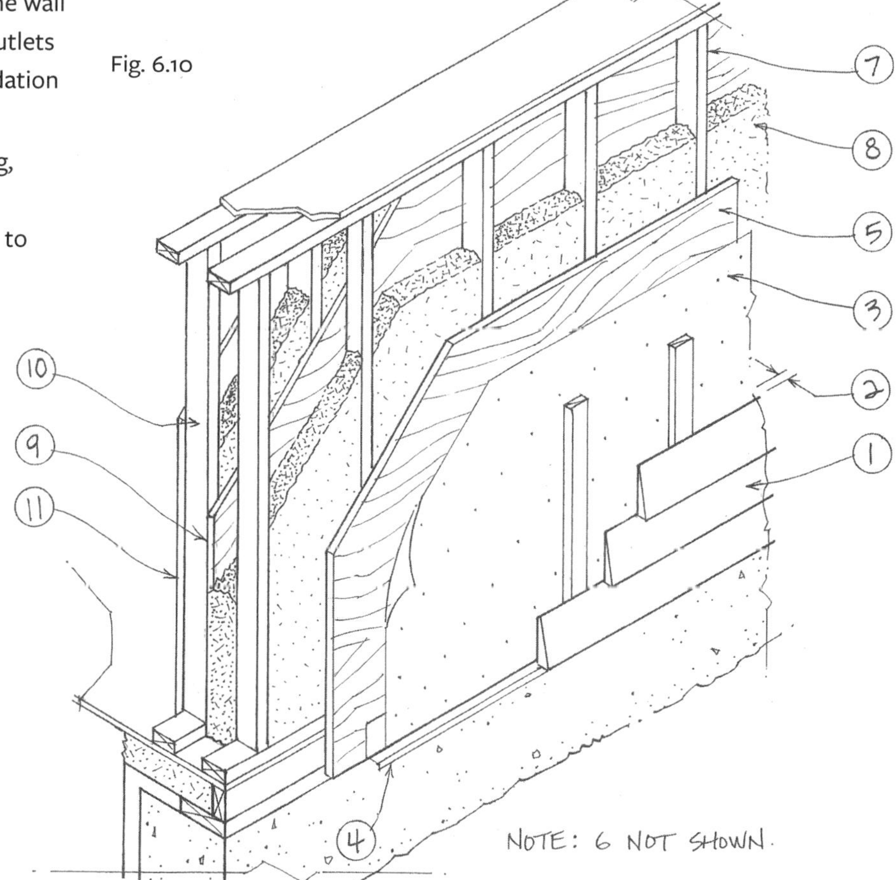

Windows and Doors

Flashing and transitions surrounding the window and door connection to walls are critical. In the "field" of a wall, it is pretty simple to get the water control layer right — just make sure the course on top laps over the course below (known as a "gravity lap" or "shingling," to support gravity-flow of water off of the wall), and provide generous overlaps at corners. Things get more complicated with windows, doors, and other transitions, however. There are a number of ways to detail window rough openings (ROs), depending on the nature of the window installation and wall assembly details.[8] Do not short-change the details, and get a second opinion before constructing if you are trying out a new, unproven method!

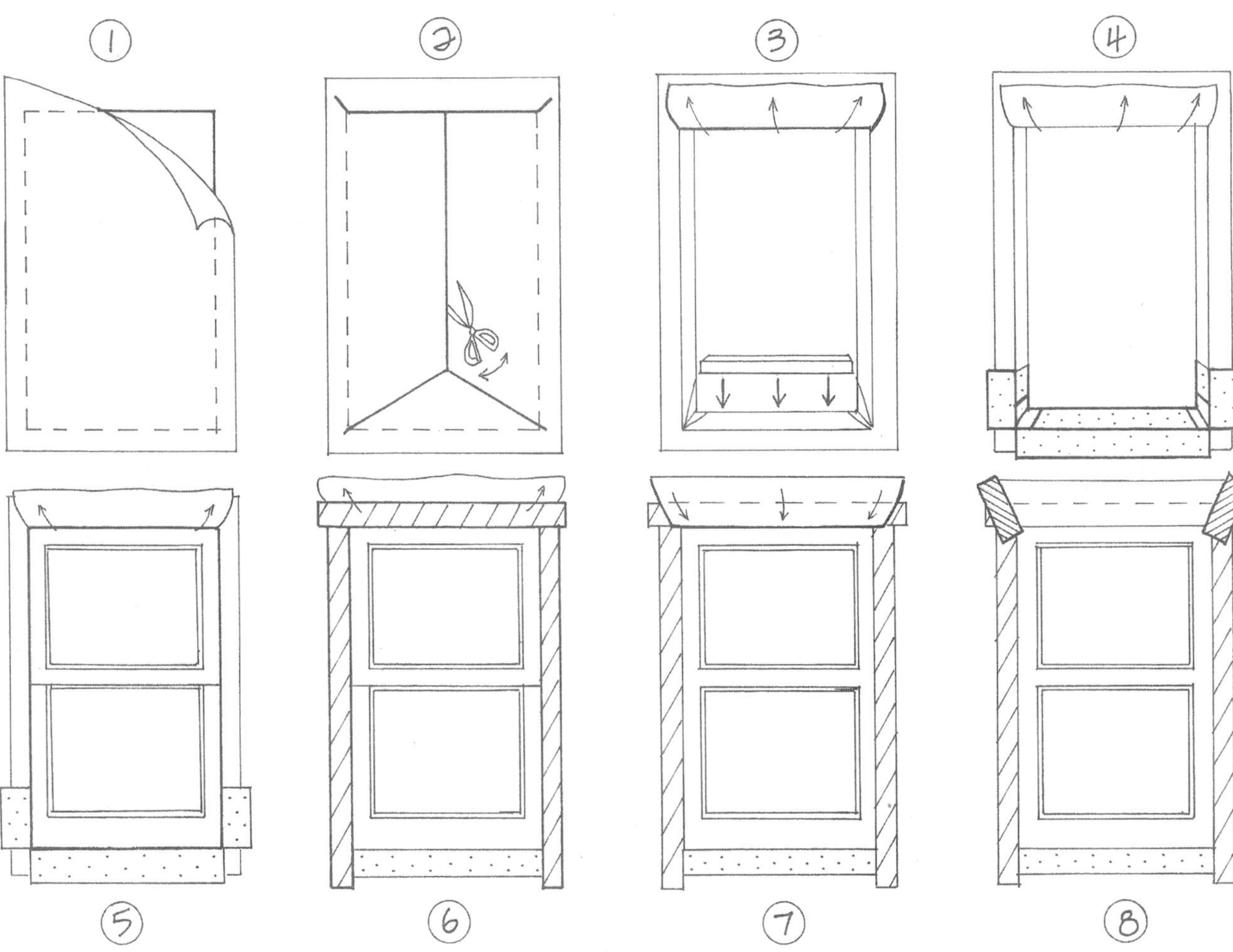

Fig. 6.11: Windows are a place where many transitions in the moisture boundary are stacked together, requiring thorough detailing to avoid premature failure and resulting damage.

1. Install WRB completely over rough opening (RO).
2. Cut an "I" in the WRB, with a horizontal cut at the top and angled cut at the bottom.
3. Fold in the sides and bottom of WRB and secure permanently, and fold up the top flap and secure temporarily; install a wood backdam and/or sloped shingle on sill.
4. Install adhesive-applied sill flashing (either a non-flexible product with corner patches, or a single-piece flexible flashing product)
5. Install window, as per manufacturer's instructions.
6. Install jamb flashing, followed by head flashing.
7. Fold down WRB at heading.
8. Apply corner patches to secure WRB head flap.

Rainscreens[9]

There is a lot of moisture vulnerability in the parts of our walls furthest to the exterior, behind the cladding. Wetting from both outside rain leakage and condensation from the inside can damage this layer, while drying potential may be limited by the cladding and lack of heat flow. To set our walls up for success, especially in wet climates or where really good drying to the exterior is critical, we can use a rainscreen system to maximize durability. In a rainscreen system, a drainage and ventilation gap is created behind the cladding and outboard of the sheathing or continuous insulation. There are several benefits to be gained from this system:

- The cladding protects the structural wall from the majority of precipitation, and protects the WRB (weather-resistant barrier) from direct wind and UV damage.
- Water that leaks behind the cladding can safely drain out of the assembly before saturating vulnerable materials further inboard.
- A capillary break is created between the cladding and the WRB, limiting wicking — this is particularly relevant for *reservoir claddings,* which hold substantial amounts of water, like unsealed brick or plaster.
- A cavity is created into which vapor from either the cladding or the sheathing can diffuse.
- If the gap is vented at both the top and the bottom, a convective air current can flow — especially on sunny walls — behind the cladding, further encouraging drying on both sides of the gap.

Dry climates or assemblies using certain vapor control strategies may not require this additional level of detailing. However, marine, wet, and cold climates with outward drying potential all benefit from this additional layer of durability. The materials used to create this gap can vary — wood or plastic strapping, 3D membranes (think of the scrubby side of a dish sponge, but as a sheet membrane), or wrinkled/dimpled housewraps are all used. While technically only a 1⁄16″ gap is necessary to provide basic function, most builders opt for 3⁄8″–3⁄4″ gaps to account for variables in construction. Horizontal strapping may be used for vertical siding (though it restricts both gravity drainage and convective drying); in extreme conditions this strapping may need to be kerfed to allow drainage. Allowing full ventilation behind the cladding maximizes drying potential; there is debate as to whether or not it is ok to vent the top of the wall into soffit/roof venting. Some experts and some codes object to the transfer of moisture from the wall into the roof assembly. Further, in regions facing brush fire risk, a formal firestop between wall venting and roof venting is required. In these cases, the wall is often vented behind a piece of trim (frieze board) at the top of the wall. In any case, any openings into the cavity must be protected from insects and rodents, either by screening or by using an off-the-shelf product designed for the purpose. Note that additional window, door, and penetration flashing must be considered and detailed as required by the variables of your wall.

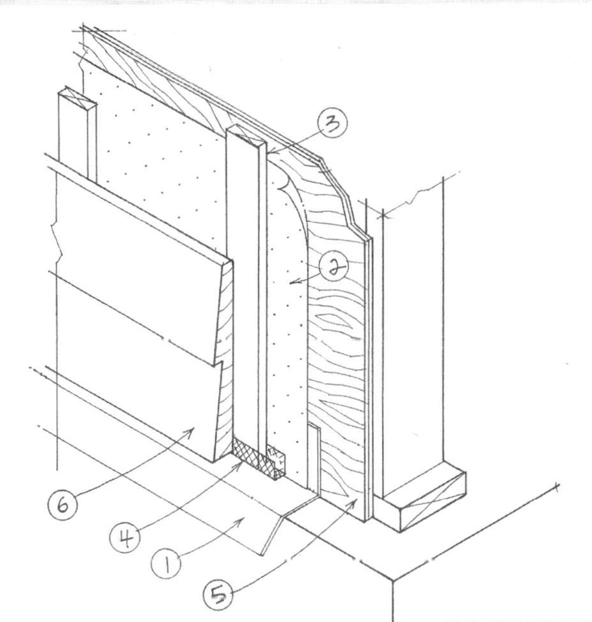

Fig. 6.12: *Rainscreen systems are valuable design tools for ensuring maximum moisture durability in wall assemblies, particularly in cold and wet regions and for well-insulated walls requiring exterior drying potential.*

1. Bottom-of-wall flashing
2. Weather-resistant barrier (water control layer)
3. Furring strips
4. Insect/rodent screening
5. Wall sheathing
6. Cladding

Roof Assemblies

In many ways, a roof is like a wall, but tilted at an angle. In roof assemblies, the water control layer, now called a *roof underlayment,* is placed over the roof decking (akin to the wall sheathing), beneath the roofing. The roofing itself (i.e., shingles, metal), like the siding, is only a first line of defense; the water control layer is a necessary redundancy to control moisture that has managed to evade your best efforts in the roofing plane, or condensation that may form beneath the roofing. The underlayment is flashed into transitions such as vent pipes, chimneys, and skylights (particularly challenging!), and the roofing is placed either directly onto the underlayment, or onto strapping to create a ventilation or drying cavity behind the roofing (more on this below). Complexities in the roof shape create valleys, hips, roof-to-wall transitions at dormers, and other situations that require additional detailing in the water control layer to ensure they stay dry under the pressure of melting snow loads and wind-driven rain. If board insulation is to be used, it is generally placed outboard of the underlayment, under strapping and/or another layer of decking, followed by the roofing. At the eaves, gutters are often employed to reduce splashback and manage ground erosion, protect entryways and open windows from the drip line, divert large amounts of water away from the foundation, and even facilitate rainwater catchment.

Given the nature of the stack effect, as well as the potential for attics to get even hotter than the outside environment, it is critical for this assembly to be optimized for performance. At a minimum, this means a high-quality air barrier; in a cold climate, you may also need vapor control. Insulation should also be carefully considered, as thermal bridges at transitions can lead to water damage problems.

Cold Attic

The simplest, least expensive, most durable, and easiest-to-build roof strategy is the cold attic. Used with gabled or other pitched roofs, an unconditioned and well-ventilated space is created between the insulation above the top-floor ceiling and the underside of the roofing. This is a case where there is separation between the layers of the assembly: the water control layer is to the outside, beneath the roofing, followed by a radiant barrier (in hot climates). On the bottom of the attic, the thermal control (insulation) is blown on top of the air control layer at the ceiling plane (vapor control may happen here too, in cold climates). *Attic slopes* are insulated rafter cavities, forming the sloped ceiling of rooms along the edges of a second floor of a gable roof building (like a Cape Cod-style roof) that run from the wall up into a cold attic. In this case, a vent space should be created in the slopes that connect eave vents in the soffit to the attic, which itself is connected to the outdoors, often through gable end vents.

 ***Advantages***

- Easily ventilated, provides plenty of outward drying potential.
- Simplified air control layer at flat ceiling.
- Access in attic allows air barrier to be inspected and maintained from inside the attic above the finished space, maintenance and servicing of radon fan, and inspection for roof leaks.
- Lots of room for super insulation; easy and inexpensive to install.
- Good design, such as a raised-heel truss, allows for continuous insulation over exterior wall assemblies (R-value over top plate should be equal to or greater than wall insulation R-value).
- Avoids ice-damming and icicles in cold climates.

- Allows for installation of radiant barriers beneath roof decking in hot climates.
- Full-height story below removes need for expensive and troublesome dormers and skylights.

Disadvantages

- Attic use for storage often compromises benefits.
- Access through interior ceiling is often a point of unwanted air, moisture, and heat transfer if not well-insulated and sealed. Exterior access is recommended for this reason (i.e., through a gable end vent on a hinge).
- Potentially raises height of building to accommodate a full story below compared to a cathedral ceiling.

From outside to inside:

1. Roofing (i.e., shingles, metal).
2. Water control layer — roofing underlayment of choice, flashed to penetrations (i.e., 30# asphalt-impregnated felt, plastic sheet membrane). Use self-sealing peel-and-stick product (i.e., Bituthene) for eave ends, valleys, and other vulnerable areas. Permeability is not relevant as it is presumed drying will occur into attic/vent space below this layer.
3. OSB, plywood, or board roof decking.
4. Radiant barrier (in hot climates).
5. Insulation wind baffle to create 2″ space, connected to continuous soffit vent and attic space (vented either by gable end or ridge vents). Material must be vapor-permeable (i.e., cardboard, plywood) if outward drying into ventilation space is desired. Baffle must be strong enough to resist insulation installation pressures (i.e., dense-pack cellulose) without crushing or deforming. Baffle must be airtight and sealed to top plate to avoid wind-washing through the insulation.
6. Open attic space, vented either by gable end or ridge vents.
7. Cavity insulation of choice on attic floor. Use vapor-permeable insulation if outward drying is desired. Note "raised heel" that lifts the rafter above the wall plate, allowing sufficient insulation above the wall top plate to reduce thermal loss and ice-damming.
8. Airtight membrane sealed to transitions and wall assembly air control layer; use a vapor control layer in cold climates when appropriate or mandated by code (Joe Lstiburek of Building Science Corporation recommends a Class I-II VR for Climate Zones 6 and 7, and a Class III VR for Climate Zone 5[10]), or select a vapor-variable membrane. Do not use a vapor control layer in warmer climates to allow sufficient inward drying.
9. Ceiling finish of choice (i.e., gypsum board, wood panels). Alternative: replace airtight membrane with gypsum board ceiling detailed as an air control layer with vapor-retarding (not vapor barrier) paint in cold climates.

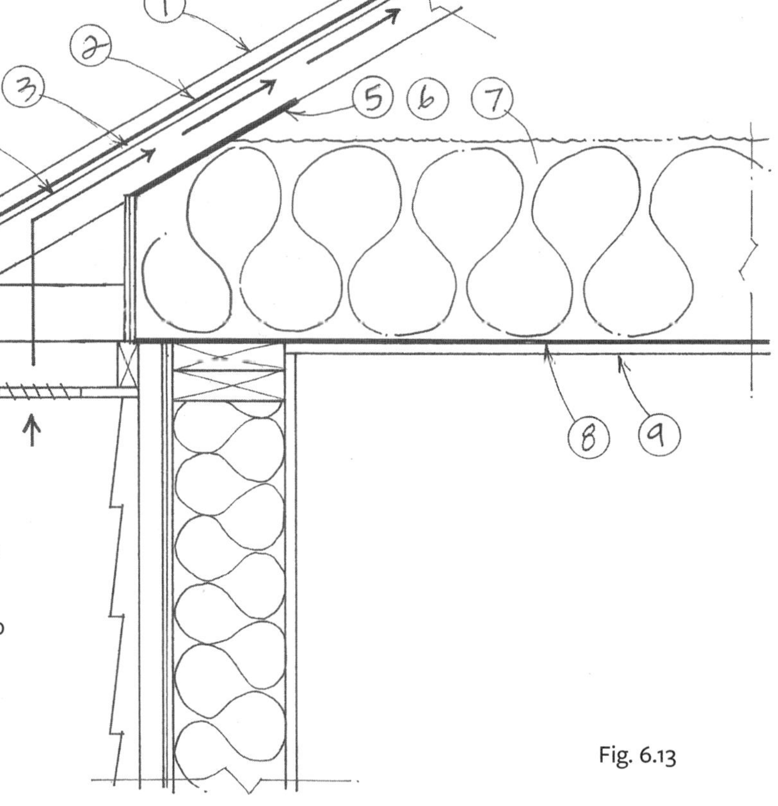

Fig. 6.13

Cathedral: Vented

Cathedral roofs are a common name given to roof assemblies in which the framed roof rafters or rafter trusses form the insulation cavity, are finished as the sloped ceiling below, and support the decking, underlayment, and roofing above. With cathedral ceilings, a vent space formed in the cavity below the decking connects to a ridge vent, where the decking is cut back to create an air space and the roof ridge material is detailed to allow air to escape but keep water from leaking in. Additional layers of continuous rigid insulation can be added below the rafters to reduce thermal bridging and increase

Figure 6.14 (insulation baffles):

From outside to inside:

1. Roofing (i.e., shingles, metal).
2. Water control layer — roofing underlayment of choice, flashed to penetrations. Permeability is not relevant as it is presumed drying will occur into vent space below this layer.
3. OSB, plywood, or board roof decking.
4. Insulation wind baffle to create 2″ space, connected to continuous soffit and ridge vents. Material must be vapor-permeable (i.e., cardboard, plywood) if outward drying into ventilation space is desired. Baffle must be strong enough to resist insulation installation pressures without crushing or deforming. Baffle must be airtight and sealed to top plate to avoid wind-washing through the insulation.
5. Cavity insulation of choice between rafters. Use vapor-permeable insulation if outward drying is desired. Confirm structural requirements for rafter sizing; increasing cavity depth will provide opportunity for additional insulation. Note construction detail that allows sufficient insulation above the wall top plate to reduce thermal loss and ice-damming.
6. Rigid insulation (optional) — if using an air-impermeable material, seal joints and transitions to penetrations and wall assembly air control layer. A vapor-impermeable material may only be suitable in cold climates.
7. Airtight membrane sealed to transitions and wall assembly air control layer, if no airtight rigid insulation above; use a vapor control layer in cold climates when appropriate or mandated by code — Joe Lstiburek of Building Science Corporation recommends a Class I-II VR for Climate Zones 6 and 7, and a Class III VR for Climate Zone 5[10], or select a vapor-variable membrane. Do not use a vapor control layer in warmer climates to allow sufficient inward drying.
8. Ceiling finish of choice (i.e., gypsum board, wood panels). Alternative: replace airtight vapor-variable membrane with gypsum board ceiling detailed as an air control layer with vapor-retarding (not vapor barrier) paint in cold climates.

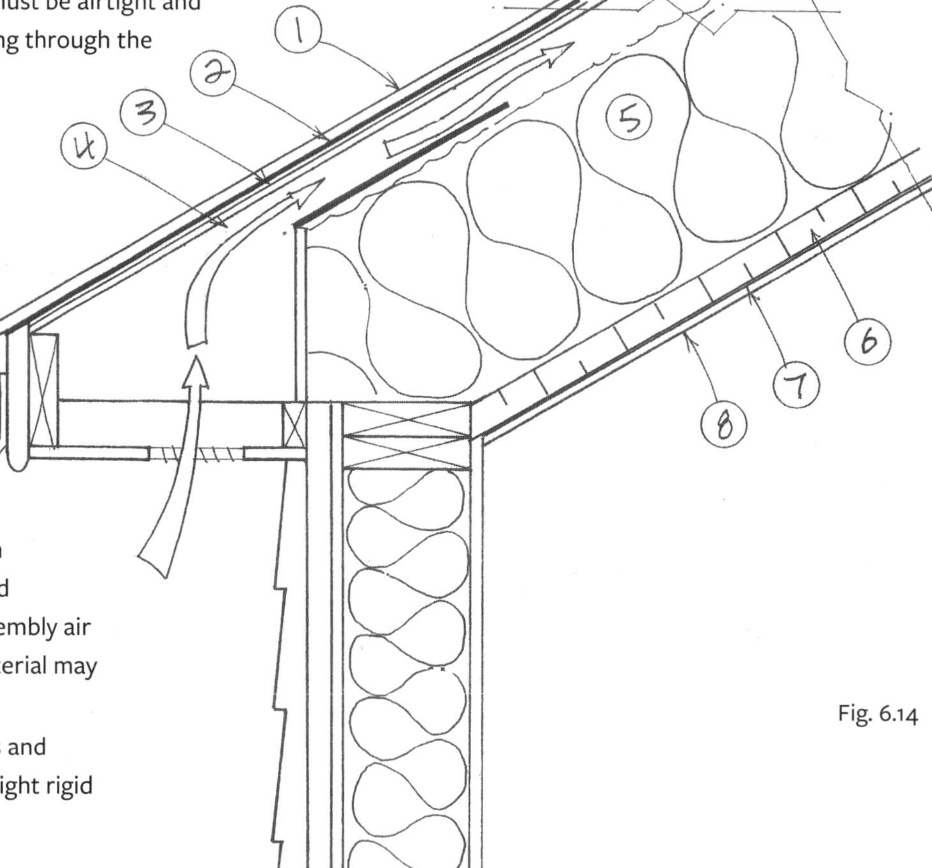

Fig. 6.14

overall R-value. An alternative approach is to fill the entire rafter cavity with insulation, and create a vent space between the decking and the roofing (sometimes a second layer of decking is required, depending on the roofing type). This allows the rafter space to be completely filled with insulation, but creates additional detailing before the roofing can go on. In this case, rigid insulation can be added either below the rafters or above the decking, with the ventilation space above the rigid insulation. See the sidebar "Vented and Unvented Roofs" for further discussion.

Figure 6.15 (ventilation above decking):

Unvented Cathedral Roof

From outside to inside:

1. Roofing (i.e., shingles, metal).
2. Additional water control layer (particularly for ice dam protection with shingle roofs).
3. OSB, plywood, board, or skip-sheathing nailing base for roofing, as required.
4. Vertical nailers to create air space, connected to continuous soffit and ridge vents.
5. Rigid board insulation (optional). Use vapor-permeable rigid insulation (i.e., mineral board) to allow outward drying of the assembly into the vent space above.
6. Water control layer — roofing underlayment of choice, flashed to penetrations. Use a vapor-permeable membrane to allow drying into the vent space above, especially in cold climates. Choose an airtight membrane and detail as an air control layer if using air-permeable board decking.
7. Plywood roof decking, seams sealed for air control, or board decking using airtight membrane above. Must use vapor-permeable material to allow outward drying; do not use OSB if outward drying is desired, and especially if rigid insulation is not included and condensation is a greater risk.
8. Cavity insulation of choice between rafters. Use vapor-permeable insulation if outward drying is desired. Confirm structural requirements for rafter sizing; increasing cavity depth will provide opportunity for additional insulation. Note construction detail that allows sufficient insulation above the wall top plate to reduce thermal loss and ice-damming.
9. Airtight membrane sealed to transitions and wall assembly air control layer; use a vapor control layer in cold climates when appropriate or mandated by code (Joe Lstiburek of Building Science Corporation recommends a Class I-II VR for Climate Zones 6 and 7, and a Class III VR for Climate Zone 5[10]), or select a vapor-variable membrane. Do not use a vapor control layer in warmer climates to allow sufficient inward drying.
10. Ceiling finish of choice (i.e., gypsum board, wood panels). Alternative: replace airtight vapor-variable membrane with gypsum board ceiling detailed as an air control layer with vapor-retarding (not vapor barrier) paint in cold climates.

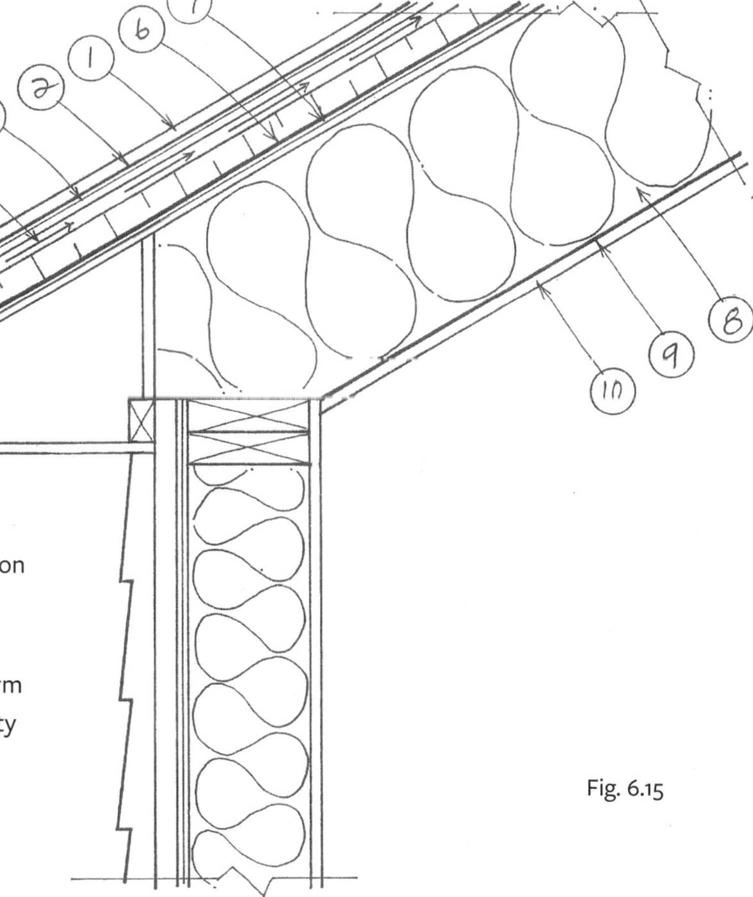

Fig. 6.15

Advantages

- Vented roof offers potential for outward drying, reducing risk of condensation damage in cold climates.
- Venting can reduce unwanted thermal transfer into the building in some cases, in hot climates.
- Venting helps avoid ice-damming and icicles, provided that appropriate thermal and air control layers are in place, particularly at wall-to-roof transition.
- Vent channels in cavities are simple to install during new construction.
- Vent channels above decking allow for maximum insulation in the cavities.
- Cathedral roofs take advantage of space below roof as high-value conditioned finished space.

Disadvantages

- Many off-the-shelf insulation baffles (which hold the insulation away from the decking) are of poor quality, and may not stand up to dense-packed or spray-applied insulation installation.
- Baffles must connect fully to soffit and ridge vents, while preventing air leakage into the insulation — details that are frequently overlooked.
- For outward drying to be effective, insulation baffles must be vapor permeable.
- Venting does not compensate for poor detailing, particularly in the thermal and air control layers.
- It can be difficult to preserve vent channels during retrofits, which in some cases can lead to moisture damage.
- Vent channels in the cavities displace valuable insulation space, and may not be practical in complex roof systems (i.e., one with valleys).
- Vent channels above the decking increase construction cost and may complicate roofing installation.
- Often, dormers or skylights are required to satisfy floor plan requirements, creating significant increases in construction cost and risk of leakage, condensation, and unwanted thermal loss or gain.
- May not provide enough insulation in cold climates without significant changes to structure (i.e., using roof trusses or wooden I-beams); use of interior rigid insulation will increase cost and may increase exposure to harmful materials (i.e., halogenated flame retardants and other chemicals in rigid foam board).

Vented and Unvented Roofs

The topic of vented versus unvented roofs is pretty complicated, and the reasoning behind making a choice will vary widely depending on your climate, roof complexity and form, assembly details, and material choices (including insulation, roofing, and underlayment). As a deep dive is beyond the scope of this book, here we will just introduce the basic concepts and define the terms.

In hot climates, the intention may be to keep the roof material cool in summer, to prolong the life of the roofing, or to reduce cooling loads.

In cold climates, the intention is often to keep the roof decking cool in the winter. This may seem counterintuitive — keep the roof cold in the winter in a cold climate? — but it reveals a common phenomenon called *ice-damming* in which excessive heat loss from either conduction, convection, or often both, warms the underside of the roof decking of a pitched roof, causing the snow pack on the roof to melt. As the meltwater runs down the roof and hits the eave overhangs, the water refreezes, and a dam of ice begins to build up. This causes water to back up, and, in the case of shingled roofs, the water can work its way back through the shingles and leak into the eave edges of the building. Cooling the decking by allowing atmospheric air to circulate above or below the decking (depending on the vent channel location) can prevent ice-damming and may also be used to allow diffusive moisture to pass out of the assembly, allowing outward drying.

The effectiveness of roof ventilation to achieve any of these purposes is highly variable. Ventilation continuity can be difficult or even impossible in complex roofs, and the stack effect is limited by roof pitch and cavity depth, among other variables. The cooling potential is quite limited in hot climates, and greater effectiveness can be achieved by changing the color of the roofing or attending to the integrity of the thermal control layer in the roof assembly. Venting is not a solution to ice-damming in the case of excessive heat loss. A quality air barrier must be created on both sides of the insulation, or venting can exacerbate convective losses. Even the drying potential will be nullified if the adjacent materials do not allow vapor to transfer through into the cavity (for example, using aluminum-faced foam board as a baffle to hold back the insulation). Unvented roofs can be designed to be quite effective — flat roofs are by nature unvented, and many quality unvented roofs have been designed by sealing the underside of the decking with foam or by layering insulation board outboard of the decking to keep the decking warm. Even foam-free unvented roofs have been successfully built using decking that can handle moderate amounts of moisture, dense-pack cellulose insulation, and a very robust interior air barrier — although it must be noted these are higher-risk assemblies and the detailing must be impeccable across the system to function correctly. Venting a roof, however, provides a layer of redundancy in a risk-prone and critical component of our building assembly, and is often worth the relatively small cost and effort required to execute in new construction.

Cathedral: Unvented

In an unvented cathedral assembly, insulation fully fills the cavity, and roofing underlayment (water control layer) and the roofing is placed directly on the decking, or directly onto rigid insulation above the decking. We can presume that there is no outward drying potential in these assemblies, and that drying must be allowed toward the interior in most cases. A very high-quality air control layer and good thermal transitions between the roof and walls are critical. Adding layers of insulation above the decking to keep the decking above dew point temperatures will greatly reduce condensation damage risk in cold climates — if the exterior insulation increases relative to the cavity insulation, as determined by the climate. Alternatively, an airtight insulation spray-applied to the underside of the decking will also protect the decking from condensation, provided that, in cold climates, the insulation is either a Class I-II VR (such as closed-cell spray foam) or has a vapor control layer applied directly below the insulation (such as open-cell spray foam with a vapor barrier paint).[11] It can be noted here that there are retrofit solutions where, when a lack of drying is creating moisture problems, a vapor-permeable membrane can replace the existing vapor-impermeable decking and underlayment at the very top of the roof, and the roofing cap reinstalled spaced off the membrane to allow an outward vapor release point at the top of the assembly.[12]

 ***Advantages***

- Simple to construct — no soffit or ridge vents, insulation baffles, etc.
- Cavities can be fully insulated; additional insulation can be added above decking without consideration of vent space.
- If detailed correctly, drying to the interior of the building can help keep moisture buildup in the assembly in check.
- Cathedral roofs take advantage of space below roof as high-value conditioned finished space.

 ***Disadvantages***

- Lack of exterior drying potential can pose a significant moisture risk, particularly with moisture-sensitive materials (see sidebar "Vented and Unvented Roofs").
- Use of hydrophobic insulation materials to seal underside of decking can result in significant moisture damage if installation is flawed or in the event of a roof leak.
- Foam insulation is often used to seal the underside of the decking to avoid condensation; reliance on foam — especially closed-cell spray foam — has significant ecological, health, and climate impacts.
- Often, dormers or skylights are required to satisfy floor plan requirements, creating significant increases in construction cost and risk of leakage, condensation, and unwanted thermal loss or gain.
- May not provide enough insulation in cold climates without significant changes to structure (i.e., using roof trusses or wooden I-beams).

1. Roofing (i.e., shingles, metal).
2. Additional water control layer (particularly for ice dam protection with shingle roofs). Note that this layer is also recommended to be a Class II VR in mixed-humid and hot-humid climates with asphalt shingles to reduce solar-powered inward vapor drive.
3. OSB, plywood, board, or skip-sheathing nailing base for roofing, as required.
4. Rigid board insulation.
5. Water control layer — roofing underlayment of choice, flashed to penetrations. Permeability is not relevant as it is presumed drying will occur inward from this layer. Choose an airtight membrane and detail as an air control layer if using air-permeable rigid insulation and/or board decking.
6. OSB or plywood roof decking, seams sealed for air control, or board decking using airtight membrane above.
7. Cavity insulation of choice between rafters. Confirm structural requirements for rafter sizing; increasing cavity depth will provide opportunity for additional insulation, but will require additional continuous insulation above to maintain decking temperatures above dew point, especially in cold climates.
8. Airtight vapor-variable membrane.
9. Ceiling finish of choice (i.e., gypsum board, wood panels). Alternative: replace airtight vapor-variable membrane with gypsum board ceiling detailed as an air control layer with vapor-retarding (not vapor barrier) paint (in cold climates).

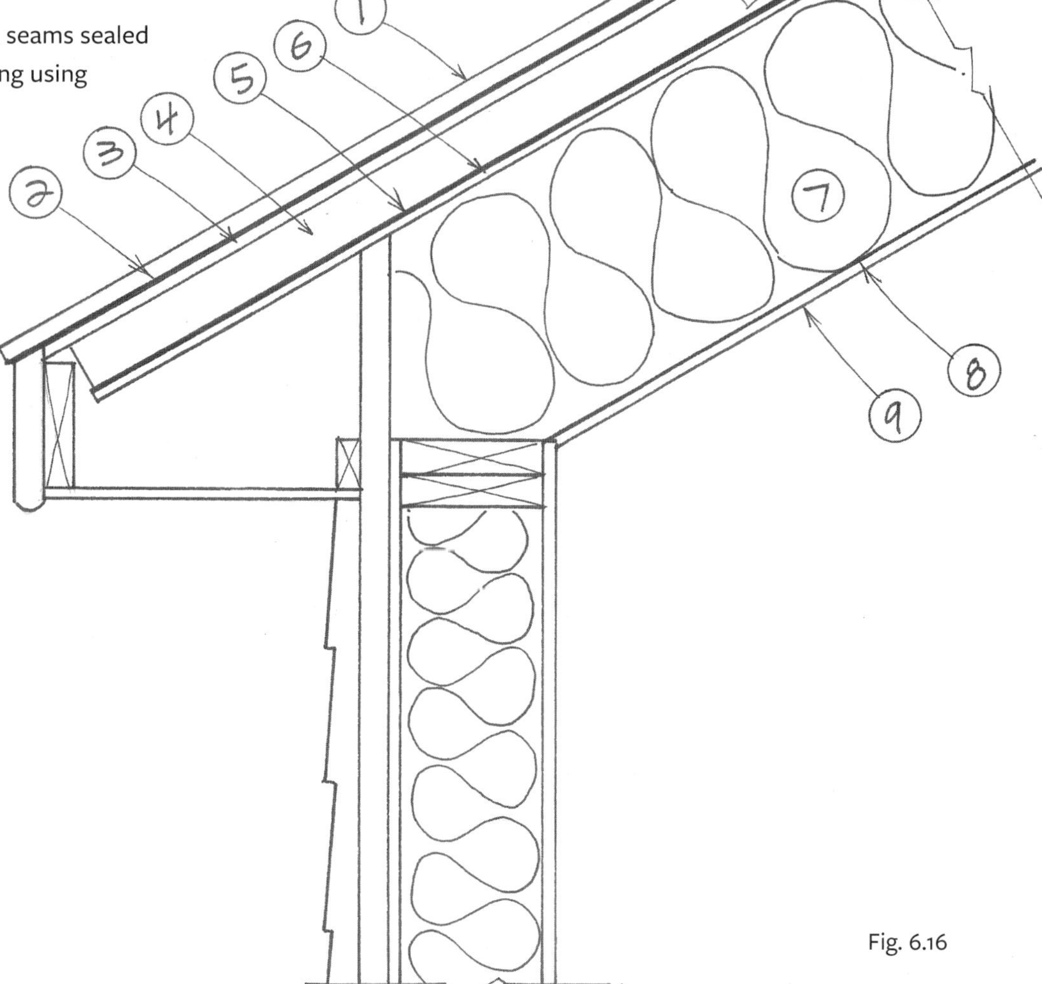

Fig. 6.16

Chapter 7

Buildings As Whole Systems

Enclosures, Systems, and Occupants

HIGH-PERFORMANCE BUILDINGS incorporate three main features: they are high-quality *enclosures* that are conditioned and outfitted with high-efficiency *systems* that are controlled and operated easily and accurately by well-informed *occupants*. All of these things must be in place; a weak link will keep the building from reaching its true potential. The mechanical, electrical, and plumbing (MEP) systems in the building must work well — under-sized or over-sized equipment will lead to discomfort, high energy costs, premature equipment failure, or even health, safety, or building durability problems. The occupants need to know how to operate the equipment correctly for it to work — a ventilation system that has not been turned on or a heat pump set to the wrong configuration will both fail in their designed roles, even if the equipment itself is correctly designed and installed. Having spent the last three chapters looking at the enclosure, we will now take a closer look at building systems and controls.

The House As a Whole System

The most important takeaway from this book is seeing our buildings as a whole system — one in which many variables interact (very much like an ecological system) rather than having distinct and separate functions. The basic description of a high-performance building displays whole-systems thinking:

- The climate in which the building is located determines the enclosure detailing and HVAC load requirements.
- As the insulation levels go up, so too must the quality of the air barrier to avoid condensation.
- A well-designed HVAC system must be tuned to the design of the enclosure and layout of the home, as well as the needs of the occupants. Ventilation is especially critical to provide, health, safety, comfort, and durability.
- Combustion appliances must have air supplies to avoid pressure dynamics on the tight home, or they should be avoided altogether.
- Occupants tend the system by operating and maintaining their equipment with clear intention, supported by adequate information from designers/builders.

It is in the relationships between the many parts of the system that the potential of the design can be realized. Systems thinking yields successful systems.

Mechanical Systems

Mechanical systems — primarily heating, ventilation, and air conditioning ("HVAC") in the case of residences — are mysterious at best, and intimidating or even terrifying at worst. This is true not only for homeowners, who frequently limit their interactions with their HVAC equipment to the dialing of a thermostat or the flick of a fan switch, but also for designers, who frequently outsource all mechanical design to mechanical engineers or builders who subcontract all mechanical system integration to mechanical contractors. Having enough grasp of the options and operations and the ability to discuss them with our professional support is important because it allows us to integrate these systems effectively and efficiently into our projects. While this is neither a complete survey of the world of HVAC nor a deep dive into the process of mechanical design, we will look at the main considerations in mechanical system selection as they relate to moisture control, thermal performance, and basic health and safety (particularly indoor air quality [IAQ]).

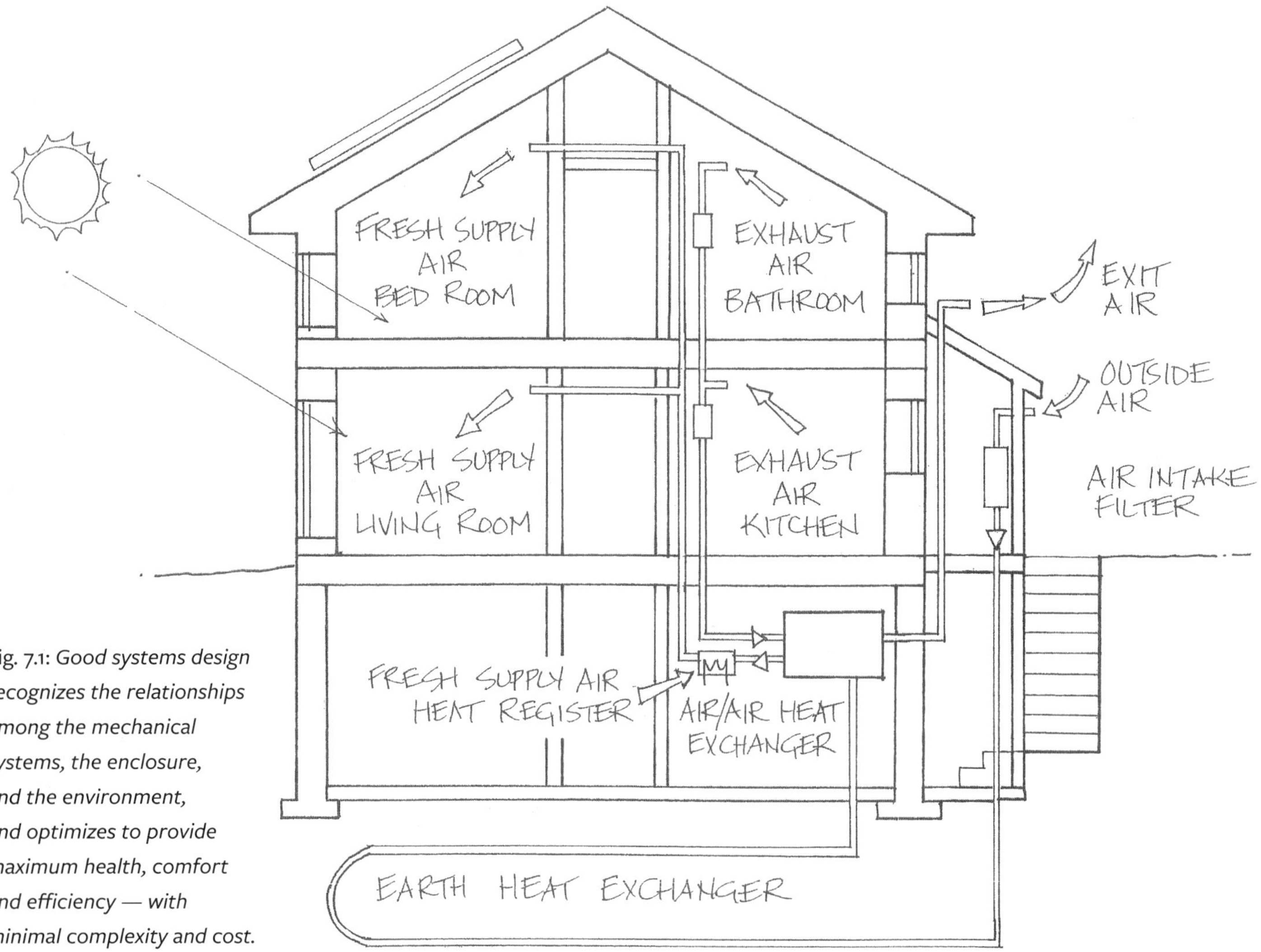

Fig. 7.1: *Good systems design recognizes the relationships among the mechanical systems, the enclosure, and the environment, and optimizes to provide maximum health, comfort and efficiency — with minimal complexity and cost.*

Heading and Cooling

Once the floor plan and enclosure have been designed, we can evaluate how to provide the basic comfort systems — heating and cooling — to the space. First, we look at *sizing* the system based on the peak loads of the building; then, we make a *fuel choice* for heating systems; finally, we decide on *system* and *distribution type*. Let's look at each of these considerations in turn:

Sizing

By entering the parameters of your design into an energy modeling or HVAC design tool, you can find the total peak heating or cooling loads of the building. Knowing how much heat is lost or gained by the building during the coldest and/or hottest times of the year gives us the information we need to accurately size the equipment to perform in extreme conditions (the winter design minimum temperature is the 97.5% worst case — to design for the one worst night of the year would be overkill). This whole-building peak load is only helpful, however, for systems that are sized to service the entire building; we still need decide how much heat or cold needs to be distributed in different parts of the house. To size space heaters or air conditioners, or to size the distribution in different rooms or parts of the house, we must conduct a room-by-room heat loss/load calculation to ensure each space in the building gets its comfort requirements met. This may seem fussy, but high-performance design requires accuracy. Oversizing a heating system by 15 kBtu is unacceptable in a building that only requires 12 kBtu on the coldest night! Many HVAC systems are oversized by 200% or more of actual load requirements, resulting in under-performing, inefficient, and short-lived equipment. Appropriate sizing will result in a more affordable, efficient, longer-lived, and lower-maintenance system.

Fuel choice (heating)

As all mechanical cooling systems run on electricity, this choice only applies to heating systems. To start the selection of your desired fuel choice, first decide whether you want combustion versus non-combustion heating appliances (see sidebar "To Burn or Not To Burn"). The lists here give some of the advantages and disadvantages of the two categories of combustion fuels:

***Fossil fuels:* including propane, fuel oil, natural gas, and kerosene**

- On-site long-term storage is relatively easy to include, but potentially hazardous (i.e., buried propane tank, basement oil tank).
- There is a simple and widespread infrastructure for delivery.
- Production and transport have very large social and ecological impacts, including the proliferation of global climate change.
- New appliances are often very efficient; sealed combustion is available.
- Systems employing these fuels are totally automated; no manual inputs required for operation, simple controls.

***Renewable fuels:* including wood, pellets, and biofuel (i.e., biodiesel, methane digester gas)**

- Solid fuels: systems are generally less automated, although automated pellet boilers and furnaces may only require occasional maintenance.
- Liquid biofuels: can be fossil fuel replacements, although the technology is still developing and socio-ecological impacts may vary widely.
- Safe long-term storage.
- Fuel is more likely to vary in quality (different grades of pellets, blends of biofuel, quality of firewood).

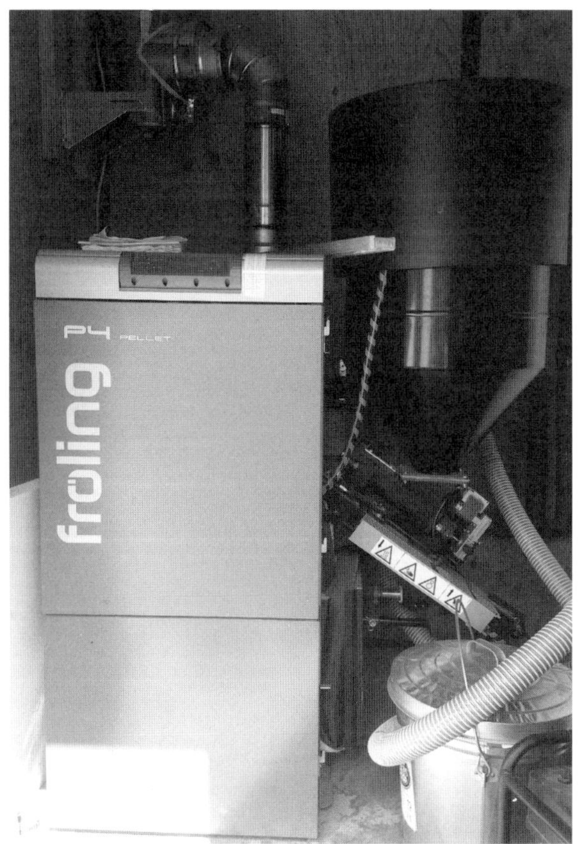

- Efficiency ranges widely, from very high for new automated pellet boilers to very low for old wood stoves burning poor fuel; accordingly, particulate and combustion gases may be higher than for fossil fuels, depending on burn quality and efficiency.
- Renewable fuel is generally not only much more socially and ecologically sound, but potentially regenerative when sourced from local, responsible providers.
- Solar energy is the most renewable and efficient of all fuels, and is a topic unto itself. In some climates, an efficient solar home can be fully heated by the sun. However, in cold and cloudy regions such as the northeastern US, a backup fuel is nearly always required to account for code or insurance requirements or weather and diurnal patterns, and solar heating (either passive solar, solar hot air, or solar hot water systems) is generally designed as an additional or "offset" fuel source.

Fig. 7.2: *Wood pellets (stored in bulk in pellet bin shown on the right of bottom photo) are an efficient and low-impact fuel in certain regions, and are now becoming as "market-stable," or more so, than fossil fuels.* CREDIT: ACE MCARLETON — NEW FRAMEWORKS NATURAL DESIGN/BUILD.

ELECTRIC POWERED APPLIANCES HAVE THEIR OWN FEATURES:

- Electric-resistance heat, while very common (especially in more moderate climates) is generally very inefficient in the big picture, as fuel is often burned to create electricity (which has inherent inefficiencies), which is then distributed (more inefficiency through transmission losses), and is then put through resistors to turn back into heat. (See the sidebar "Primary Energy vs. Secondary Energy").
- Heat pumps — either air-source or ground-source — have higher efficiencies; they produce 3–5 times the amount of energy (as heat) as electricity required to run them.
- Non-combustion heat is a big plus for IAQ — no combustion gases, no particulates, no potential back-drafts from building pressure dynamics.
- Minimal penetrations in the enclosure — no chimney flues or make-up air ducts to detail.
- No on-site fuel storage risk or space requirements.
- Potentially less resilient in the face of storms or power disruptions, depending on grid reliability, unless backup power provisions are made (i.e., stand-by generator, battery storage).
- Potential for PV/renewable offset for net zero construction.

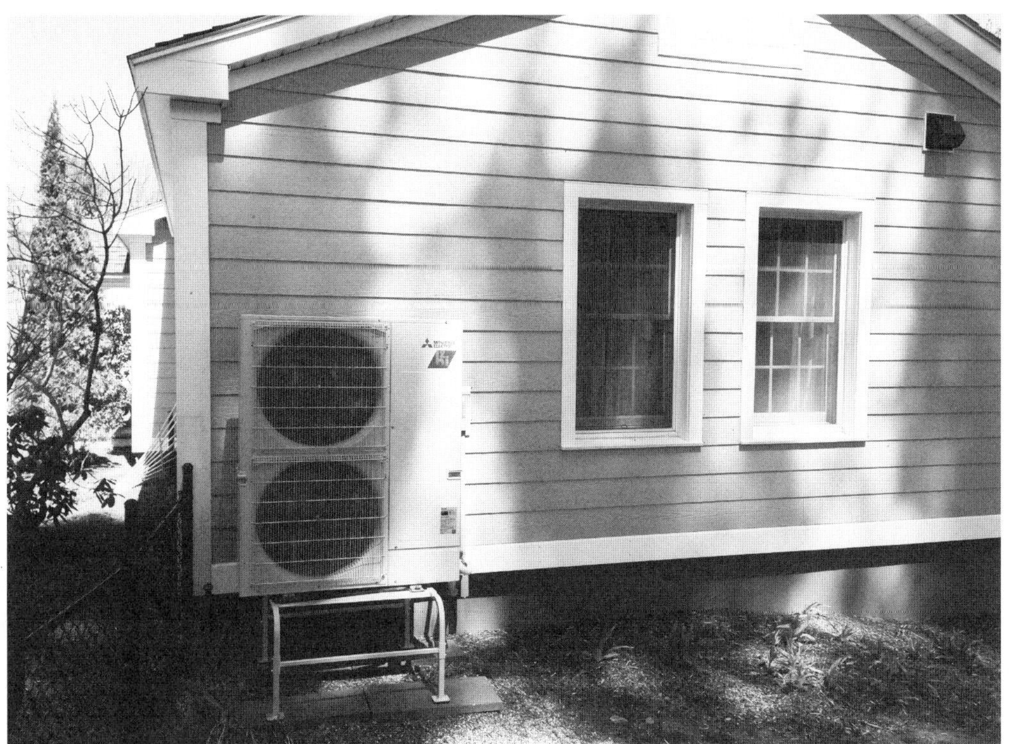

Fig. 7.3: *Mini-split air-source heat pumps provide both heating and cooling in a wide range of climates very efficiently, relying on electricity for power.*
CREDIT: JACOB DEVA RACUSIN — NEW FRAMEWORKS NATURAL DESIGN/BUILD.

To Burn or Not To Burn

One of the biggest questions in assessing the energy needs of a high-performance building is whether or not to incorporate combustion equipment into the design. Many high-performance homes feature all-electric appliances, for a variety of reasons:

- Tight enclosures may make it easier for combustion appliances to have exhaust gases sucked back into the building rather than leaving through the chimney (back-draft) when a ventilation exhaust fan is operating.
- IAQ impact of combustion appliances may be undesirable (especially wood-burning appliances).
- It may be harder to find combustion appliances that can work with the small heating demand of a low-load building.
- Gas and particulate emissions may be avoidable altogether by using an electric system offset by solar electric (PV) panels.

However, combustion appliances can certainly be successfully integrated, and there may be plenty of reasons to prefer them: there may be no good non-combustion mechanical options; the building may have limited electrical service (i.e., an off-grid building); or a biofuel may be less of an ecological impact than electricity from a polluting source (such as a coal plant), or it may provide greater resilience. To that end, a few considerations are very important[1]:

- Ensure you use sealed combustion appliances that source dedicated combustion air from the outside that is ducted directly to the appliance; this isolates the appliance from the atmosphere of the building, ensuring the appliance will always get the air it needs for proper combustion, and that pressure dynamics in the house (for example, if a large exhaust fan is turned on) won't cause the appliance to back-draft. Wood stoves should be new, airtight, and have a dedicated air supply to the firebox.
- If there is a gas cooktop or oven, make sure a range hood is properly installed and vented to the outdoors.
- Install a hard-wired carbon monoxide (CO) alarm on each floor, and near each bedroom.
- Ensure a qualified service technician provides regular maintenance on the appliance, with a service record for reference.

Primary Energy vs. Secondary Energy

When most of us think of the energy we consume in a building, we are thinking of *site energy* — that which is reflected in our utility and fuel bills. This energy is delivered to our homes in one of two ways. *Primary energy* is the raw fuel that is burned on site, such as natural gas or fuel oil. *Secondary energy* is the energy product created from a raw fuel, such as electricity from the grid (created by a variety of different fuels) or heat from a steam system. Therefore, a unit of heat created from a wood stove and a unit of heat from an electric heater are not directly comparable, since the heat from the electricity, as a converted fuel, also embodies the raw fuel used to create the electricity at the power plant, as well as the distribution losses. To fully account for all of the energy inputs in our buildings, it is important to look at *source energy,* which is a combination of primary and secondary energy. Table 7.1 shows the *source-site ratio* that can be used to account for the primary energy of any given fuel.[2] For example, 10,000 BTUs of heat provided by grid electricity in the United States will, on average, have a source energy load of 31,400 BTUs. Those extra BTUs come from all of the inefficiencies ⟶

incurred by burning fuel in the plant and distributing it to your home. This is why electric-resistance heat is considered to be so inefficient (unless it is powered by renewable energy!). The embodied energy of our primary fuels has a very real impact on the global warming potential of our homes. If ecological impact is a priority, fuel choice must be carefully considered.

Table 7.1: Source-Site Ratios[3]

Energy Type	Source: Site Ratio
Electricity (national grid average)	3.37 in US; 2.05 in Canada*
Electricity (solar PV)	0.2**
Natural Gas	1.1
Fuel Oil (1, 2, 4, 5, 6, Diesel, Kerosene)	1.18
Propane and Liquid Propane	1.15
Kerosene	1.21

* Source: "EnergyStar PortfolioManager Technical Reference", USDOE EnergyStar, portfoliomanager.energystar.gov/

**Reflects embodied energy of PV panels relative to electricity output. Source: Raugei, M., Fullana-i-Palmer, P., and Fthenakis, V., "The Energy Return on Investment (EROI) of Photovoltaics: Methodology and Comparisons with Fossil Fuel Live Cycles," Energy Policy Vol 45, June 2012

Fig. 7.4: *Although we account only for site energy when we look at our utility bills, the source energy includes all the wasted and consumed energy of production and distribution — the "embodied energy" of energy.*

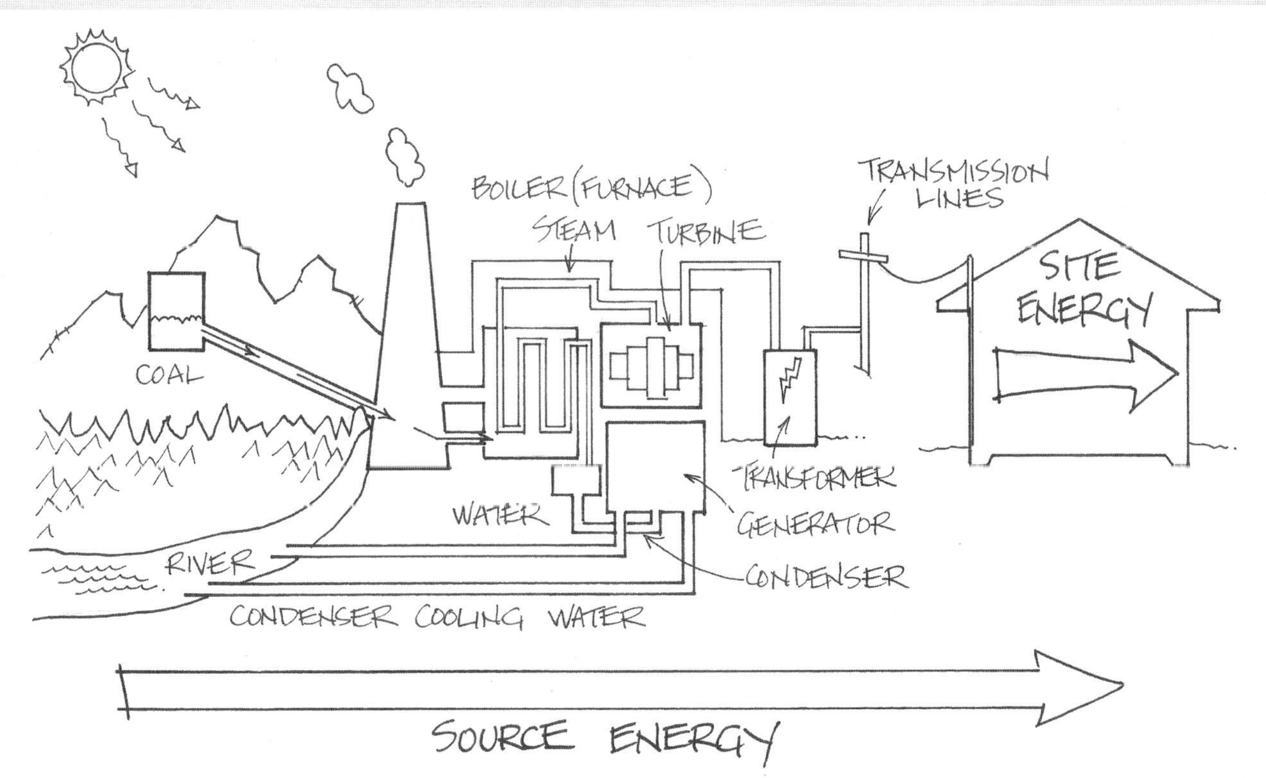

System Type

There are many options in different heating or cooling systems, but there are two basic approaches: you can heat and cool fluids in a given *space*; or you can heat and cool fluids in a central piece of equipment, which are then *distributed.*

Space heating or cooling requires a point-of-use unit that conditions the immediate space in which it is located. If your building has a high-quality thermal enclosure, space conditioning may be all that is necessary to keep the building comfortable.

A distributed system (also known as *central heating or cooling*) employs a larger piece of heating and/or cooling equipment in a central location. Hot or cool fluids (air, water, coolant) are distributed to different parts of the building by way of ducts or pipes. There are two main choices for distribution. In ventilation systems (discussed later) hot or cold air runs through ducts (sometimes these are *forced-air* systems). In liquid, or hydronic, systems, hot (and sometimes cold) fluid is pumped through pipes attached to radiators of various designs.

There are many high-performance, renewable, and fossil-fuel-free solutions on the market — with more emerging every day. Many of these rely on the integration of multiple systems, such as solar hot water systems that "preheat" a boiler system. Ultra high-performance systems that feature integrated heat recovery ventilation and ducted heat pumps, long available in Europe, are starting to appear in North America. With time, more innovation and market acceptance will increase the options available to us.

Fig. 7.5: *Two strategies for heating: central heat distributed throughout the building, or space heat placed in strategic locations.*

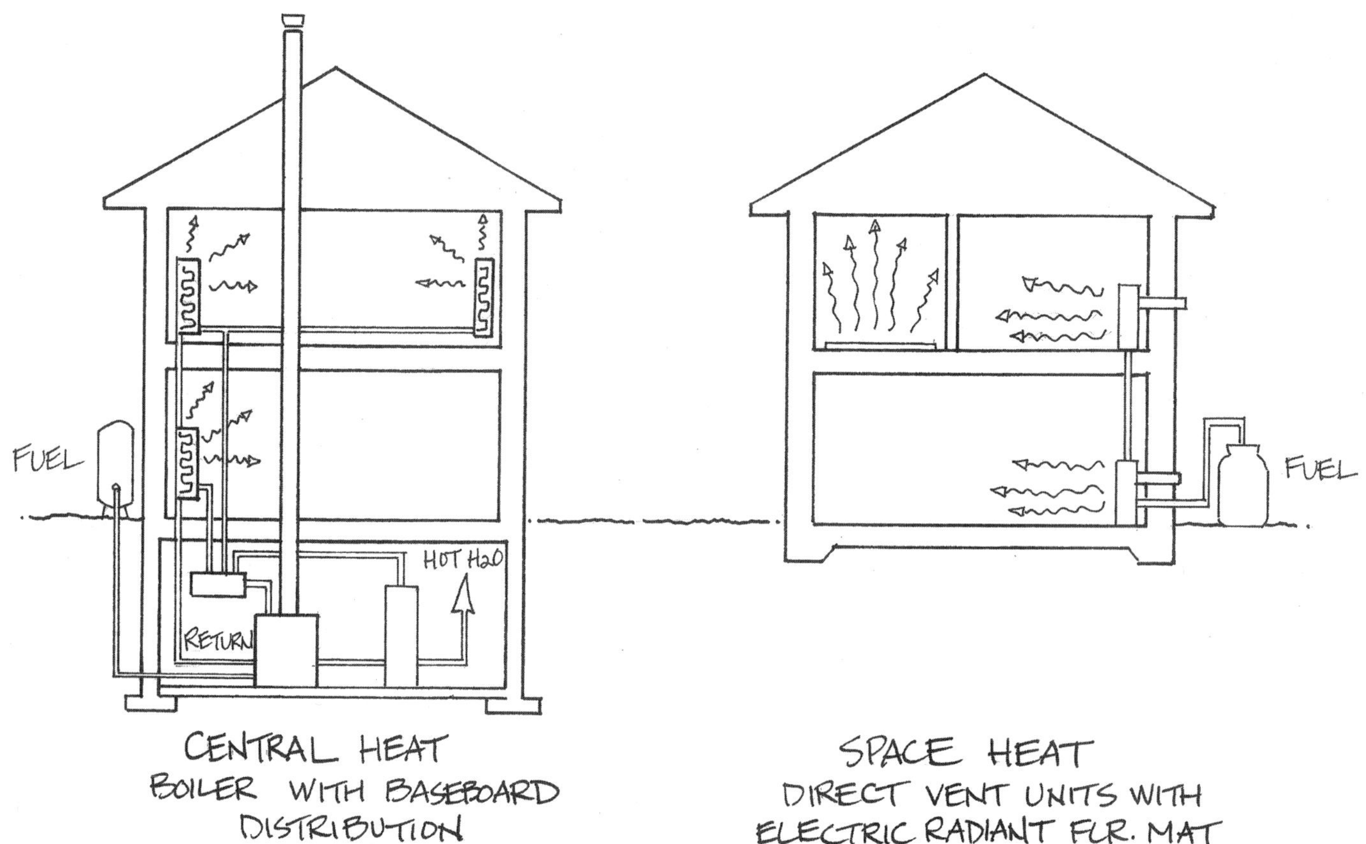

Table 7.2: System Type

Criteria	Space Conditioning	Distributed System
Capacity	• lower capacity (10–75 kBTU) • better for tight buildings or small spaces	• higher capacity (50–150 kBtU) • better for larger or leakier buildings
Control	• more smaller units offer good local control • less convenient to operate than a single central plant	• buildings can be divided into "zones" for control • spaces are easily controlled with programmable or newer "learning" thermostats
Cost	• cheaper installed cost, unless multiple units are needed • low ongoing maintenance costs	• cost is usually higher, both for appliance and distribution • ongoing energy and maintenance costs to run blower motors, circulation pumps, etc.
Comfort	• comfort issues when located in a sleeping or living space from turbulence (blowing air) or noise (i.e. from fans)	• comfort issues can arise from ducted systems: turbulence, noise, dust from improperly maintained systems • hydronic systems are valued for being quiet and non-turbulent
Efficiency	• high efficiencies can be realized from appropriate sizing • high-efficiency units commonly available	• requires finding appliances that can "turn down" to efficiently operate at lower loads • high-efficiency units commonly available • distribution losses must be considered
Design Flexibility	• difficult to integrate with other systems • greatest flexibility from using multiple units	• can integrate multiple systems (i.e. forced-air heating, cooling, and ventilation or supplying DHW from boiler) • easier to integrate multiple fuels (i.e. solar, fossil, wood coil)

Table 7.3: Fuel Type

Fuel Type	Hydronic	Air		Other
	Distributed	Space	Distributed	Space
Electric (heating)	• electric tank/boiler (smaller loads) • ground-source heat pumps • air-to-water heat pumps (limited options)	• blower-assisted resistance heaters • ductless air-source heat pumps*	• ducted air-source heat pumps • ground-source heat pumps	• resistance lamps/ panels/floors
Electric (cooling)	• chilled beams, radiant cooling (commercial)	• window unit ac • ducted evaporative or *swamp* coolers** • ductless air-source heat pumps	• central ac • ducted evaporative or *swamp* coolers** • ductless air-source heat pumps • ground-source heat pumps	
Fossil (heating)	• boiler supplying baseboard, radiator, radiant floors, ceilings, panels; can also supply DHW • tank heater w/ heat exchanger (for smaller loads)	• direct-vent or chimney heater (optional blower)	• forced-air furnace	
Biomass (heating)	• pellet or wood boiler supplying baseboard, radiator, radiant floors, ceilings, panels; can also supply DHW	• pellet or wood stove, wood fireplace, wood masonry heater	• pellet or wood forced-air furnace	
Other Renewable (heating)	• solar hot water, integrated to backup hydronic system	• solar hot air space heaters		• passive solar heating (enclosure design)

* Ductless air-source heat pumps, or mini-splits, provide both heating and cooling at far greater efficiencies (250%–500% greater than standard electric units); cold-climate units can operate reliably in heating mode at temperatures as low as -15F (-26C).

** For use in hot dry climates, in which air is blown through a wet membrane, requiring additional heat from the interior to induce phase change and evaporate the moisture. "Swamp" refers to any evaporative cooler, ducted or ductless.

Ventilation

Ventilation is a mission-critical requirement for any building. In addition to providing comfort, appropriate ventilation is necessary to ensure the health and safety of the occupants, the durability of the building, and the appropriate functioning of the rest of the mechanical system; it can be designed to also contribute to energy efficiency while providing these services. Code-required mechanical ventilation is often limited to bathroom exhaust fans and a kitchen exhaust hood, but high-performance goals dictate a more thorough approach.

As mentioned earlier, leaky buildings do not ensure the quality of the indoor air (in fact, it is often quite the opposite, as the "fresh air" is being pulled in through leaks in crawl spaces, garages, and dirty walls); nor do they ensure that the correct amount of air is reaching the right rooms at the right times.

Build It Tight, Ventilate Right — But How Much?

The adage "build it tight, ventilate right" refers to the importance of limiting air leakage through our assemblies, and then ensuring a good ventilation strategy is in place. "Ventilate right" covers both the correct amount of ventilation and appropriate circulation throughout the different parts of the building (i.e., supply in bedrooms, exhaust in bathrooms). Answering the question "how much?" is more complicated than it may seem. ASHRAE Standard 62.2 is the ventilation standard that most codes and common practices reference for residential construction (ASHRAE stands for American Society of Heating, Refrigeration, and Air Conditioning Engineers). Table 7.4 shows calculated values from the ASHRAE Standard for continuous supply.[4]

In high-performance buildings, this standard may result in over-ventilation, which can result in energy waste, discomfort, and even moisture problems in humid environments when too much humid air is introduced for the low-load cooling system to handle, requiring additional dehumidification. In buildings where there are lots of pollutants (i.e., smoking, synthetic finishes), this standard may not be enough.

So how much ventilation does the building need? That depends on many factors, including the performance and materials of the building, what is inside of it, and where it is located. The goals of the ventilation system — for both spot and whole-house ventilation — are to remove pollutants when possible, dilute pollutants when not, ensure appropriate circulation throughout the building, and help manage

Table 7.4: Whole Building Ventilation

ASHRAE 62.2-2013 — Whole Building Ventilation Requirements					
Number of Bedrooms	1	2	3	4	5
Area (SF)	*CFM*	*CFM*	*CFM*	*CFM*	*CFM*
1000	45	52.5	60	67.5	75
1500	60	67.5	75	82.5	90
2000	75	82.5	90	97.5	105
2500	90	97.5	105	112.5	120
3000	105	112.5	120	127.5	135
3500	120	127.5	135	142.5	150

humidity to safe and comfortable levels. To achieve this, start by sizing the ventilation system according to the ASHRAE Standard 62.2, but then be prepared to adjust the equipment (with the help of a professional) if the building seems over- or under-ventilated. One way to quantify this is the interior relative humidity; the Building Science Corporation offers the following recommendations,[5] which should be balanced against optimization for human health (see Figure 7.6):

- Between 20% (during winter) and 60% (during summer)
- No more than 35% during winter in Zone 5
- No more than 30% during winter in Zone 6
- No more than 25% during winter in Zone 7
- No more than 70% for a few days in air-conditioned homes

Types of Ventilation

There are two types of ventilation: *spot ventilation* and *whole-house ventilation.*

Spot ventilation is required in specific locations to remove humidity and pollutants from a building. Common locations are kitchens, bathrooms, and clothes washing areas, but dedicated vents may be needed for workshops, root cellars, pool rooms, attached greenhouses, or other special use cases. These vents MUST be ducted to the outside. Some particularly high-powered fans, such as large range hoods (higher than 150 CFM), may also require make-up air via dedicated air intake ports. Large range hoods are notorious for creating significant depressurization conditions in tight buildings, and must be designed accordingly (especially if atmospherically vented combustion appliances are present in the building).

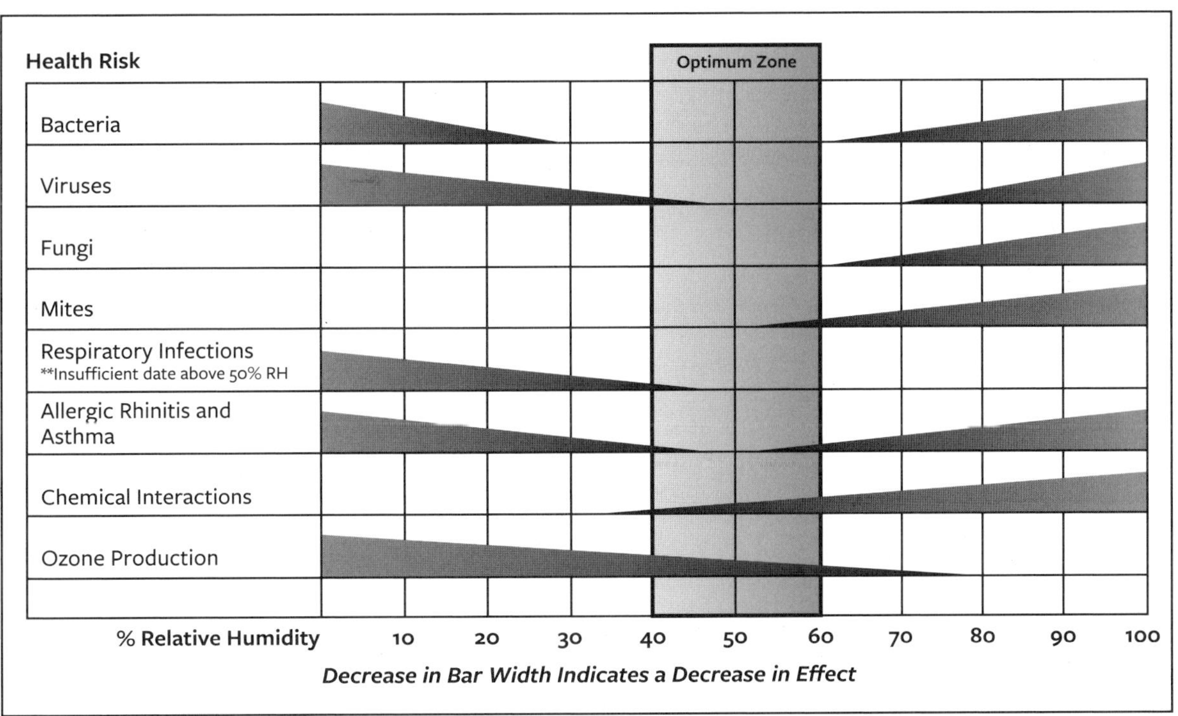

Fig. 7.6: *Optimum Relative Humidity for Health — The Effect Indoor Humidity has on Health Risks.*[6]

Whole-house, or dilution ventilation, manages pollutant and relative humidity concentrations, provides circulation throughout the building, and supports a healthy and comfortable environment. There are three main types of mechanical dilution ventilation: supply-only, exhaust-only, and balanced.

Finally, it should be noted that for all ventilation systems, as for forced-air systems, the quality of the system relies upon the quality of the ductwork. Appropriately sized solid-wall ducts with minimal bends and short runs are required to ensure the performance of the system; these must, of course, be vented to the exterior.

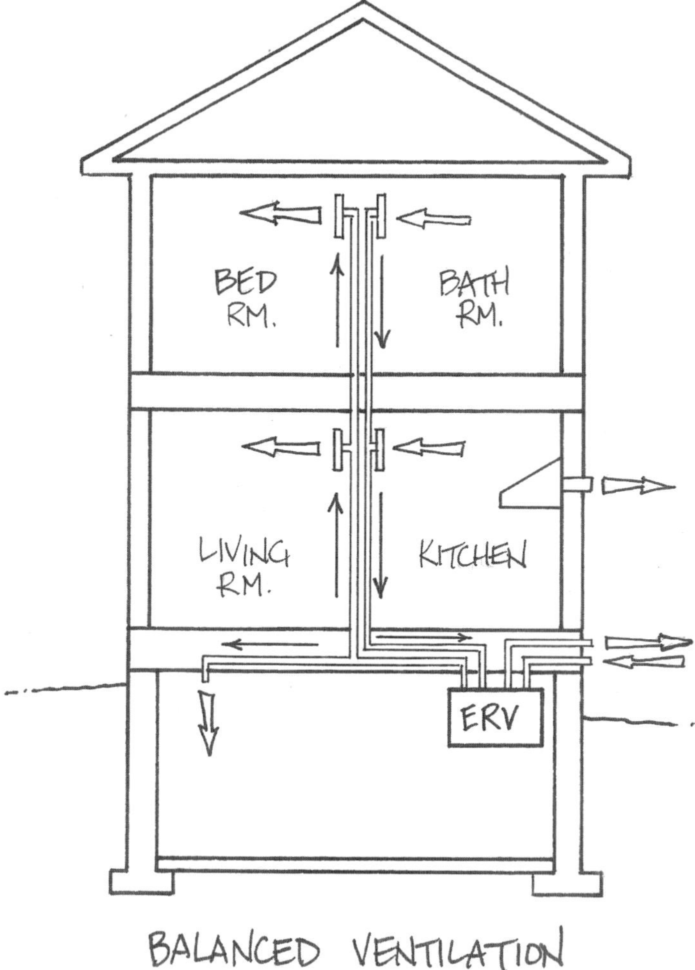

Fig. 7.7: *Balanced, exhaust-only, and supply-only ventilation systems with spot exhaust in bathrooms and kitchen.*

Let's take a look at the functions and features of the various dilution ventilation system options:

Balanced (HRV, ERV)

In a balanced ventilation system, both exhaust and supply are handled by the same system. These systems feature the added benefit of being able to handle heat and moisture exchange, dramatically increasing both the comfort and efficiency of a system. There are two basic types of systems: *Heat Recovery Ventilators (HRVs)* and *Energy Recovery Ventilators (ERVs).* In both, supply and exhaust ducts to the outdoors are connected to the system, which then lead to supply and exhaust ductwork within the building; they are powered by low-energy, low-noise, continuous-use-rated fans. With an HRV, the exhaust and supply streams pass by each other in an air-to-air heat exchanger. With an ERV, the air streams are forced through a membrane, which acts as both a heat and a humidity exchanger. The "energy" referred to in ERV is the latent energy of water. As we discussed in Chapter 3, there is a lot of latent energy in water vapor, which can add considerably to a building's cooling load. In a hot-humid climate, rather than outside humidity passing into an air-conditioned building and adding to the cooling load, it is sent back outside; thus, efficiencies for ERVs can be as high as 90%. ERVs tend to be more popular in hot climates for that reason, although they may also be used successfully in cold regions, providing that excessive indoor humidity isn't a problem in the winter (HRV ventilation can provide some desirable dehumidifying benefits). A few types of ductless units are now available as well; these rely on passive mixing within the building (no ducts) but still offer balanced pressure and heat exchange.

Exhaust-only

Perhaps the least expensive and most common in cold climates, this system relies on a fan ducted to the outdoors to blow air out of the building. The effective use of exhaust-only systems is a subject of debate among experts, because the building is depressurized as stale air is exhausted, forcing make-up air to leak in through gaps in the enclosure. While the depressurization of the building is helpful for reducing condensation in cold climates, it can increase risk of pollutant infiltration (such as humidity, mold, dust, radon, or other chemicals when make-up air is pulled through a basement or garage), or noxious gases if the fan contributes to the back-drafting of combustion appliances. These problems can be mitigated by including passive make-up air inlets into the design of the house. Inlets can be outfitted with filters, dampers or controls and integrated in strategic locations to encourage circulation and targeted supply. However, they are not reliable unless the enclosure is very tight; they are often overpowered by wind forces, and many of them may be required to adequately balance the exhaust system.

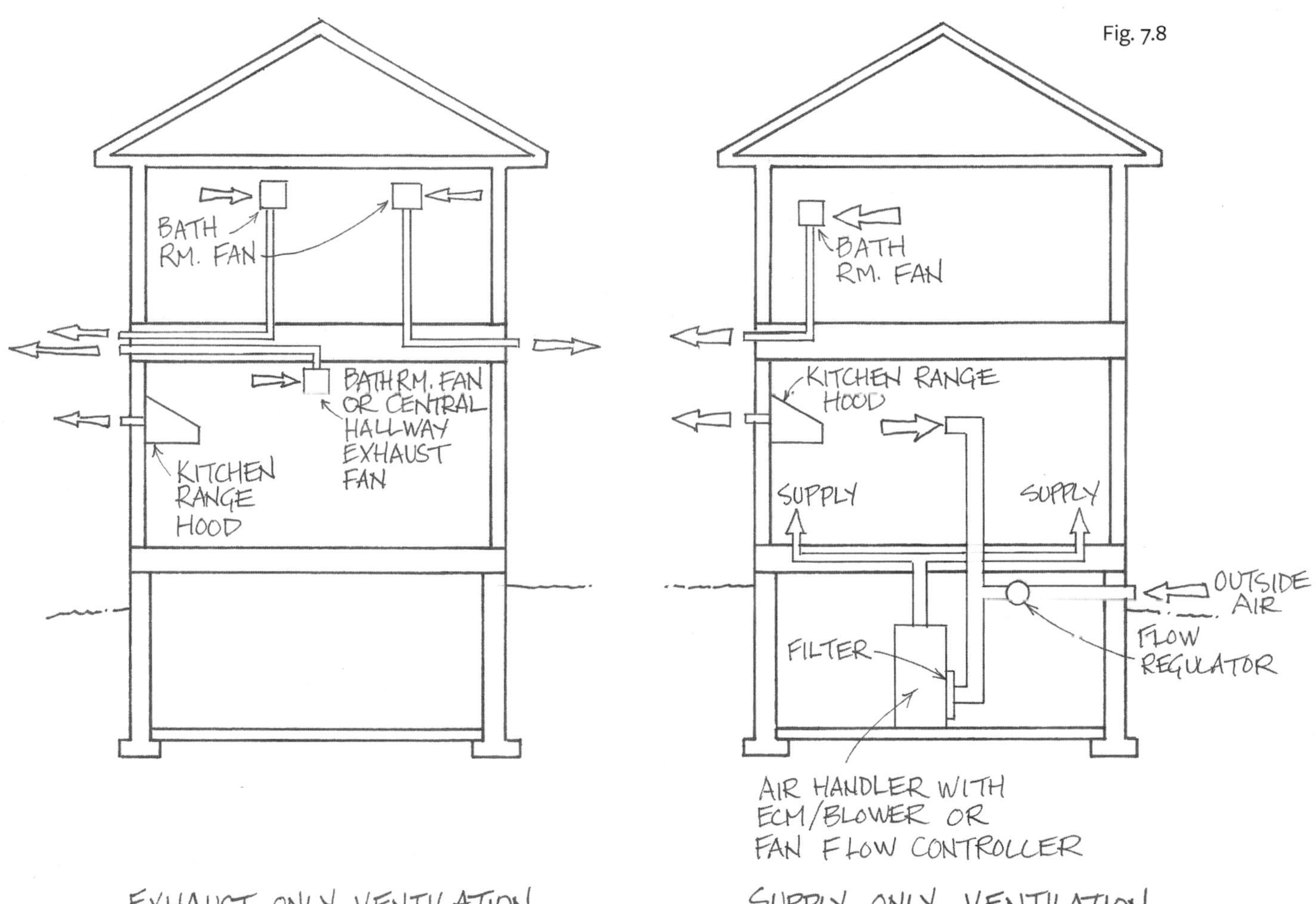

Fig. 7.8

Table 7.5: Ventilation

Criteria	Balanced (HRV/ERV)	Exhaust-only	Supply-only
Pressure Balance	• pressure balance is normalized by controlled air intake and exhaust; no pressure balance issues	• relies on enclosure leaks and depressurization for supply distribution or make-up air inlets • especially problematic for buildings with radon, atmospherically vented combustion appliances (unless make-up air inlets are effectively included in the system)	• relies on enclosure leaks and pressurization for exhaust, supply is commonly incorporated into a forced-air ducted heating/cooling system
Control	• appropriate levels of partially conditioned fresh air are supplied when and where needed (bedrooms and living rooms) • exhaust air removed from bathrooms, kitchens, and basements (spot ventilation may still be required)	• supply distribution location and amount is largely uncontrollable (except when make-up inlets are used) • requires automated or programmable controls (i.e., for humidity, CO_2, motion sensor) to ensure appropriate ventilation	• supply air is ducted to desired locations • requires controlled electronic damper and mechanical balancing damper to regulate air flow; frequently omitted leading to poor ventilation rate control
Cost	• most expensive systems to install (need to be installed, balanced, and commissioned by professionals) • ductwork increases cost, especially in a retrofit • maintenance is simple as regular filter cleaning, but is critical	• least expensive systems to install (cheap fan, minimal or dual-purpose ductwork)	• cost of most ductwork covered by forced-air system, otherwise very expensive • supply-only systems must be regularly maintained by a professional
Comfort and Health	• best choice for health and safety • heat exchange conditions fresh air supply • ERVs manage moisture exchange as well • noise and turbulence may be issues; require care in design • regular filter changes required	• little option for heat or moisture exchange • noise from fan may be a concern • depressurization may lead to health and safety issues (i.e. soil gases, combustion appliances) • no filtration option unless make-up inlets have integral filters	• temperature control may be handled by forced-air unit • noise and turbulence are issues with many forced-air systems • condensation in ducts a concern in humid climates • regular filter changes required
Efficiency	• energy and heat recovery boosts efficiency of the units • appropriate ductwork (smooth, sealed, well-sized, minimal bends and runs) is critical • energy-efficient, variable-speed fan motors are available	• little option for heat or moisture exchange • energy-efficient units are available • often least expensive to operate and no maintenance required	• little option for heat or moisture exchange • variable-speed air handlers required to avoid using big fans to move small air flows • often requires mechanically operated dampers
Design Flexibility	• running ducting in retrofits is challenging • consider ductwork locations early during the floor plan development • ductless units on the market, including electronically controlled paired fans and "pipe-in-pipe" concentrically vented units	• easiest to integrate into a retrofit	• integrates into forced-air system; relies on existing design

Supply-Only

In this system, air is pulled from outdoors into the building; this is powered and distributed by a forced-air system (and therefore more common in mixed and hot climates). This pressurizes the building, which forces exhaust air out through leaks in the building (less of a durability concern in warmer climates, where condensation risk is lower). A flow regulator (often a motorized damper) controls the amount of incoming air to avoid over-ventilation, which can lead to energy losses and humidity problems — this last detail is often overlooked, but it is critical to avoid under- or over-ventilating the building. Condensation in the ductwork can be an issue in humid climates, and variable-speed motors should be used to efficiently move the relatively small amounts of ventilation air when there is no heating/cooling demand.

Passive Ventilation

It is a common myth that passive, or non-mechanical, ventilation — supplied by leaks in the enclosure and operable windows — is a satisfactory strategy. Building a leakier house will not lead to an appropriate exchange of air.

Passive ventilation is driven by two physical phenomena: the *stack effect,* which is a factor of a difference in temperature (ΔT) between the inside and outside of the building, and wind pressure. If there is very little ΔT and very little wind — think of a still, mild day — there will be very little passive ventilation potential. Even if infiltration supplies air into the building, it is often contaminated because it is filtered through musty basements, attached garages, and wall cavities that contain mold and dust. Further, if the building is tight and you have opted not to open your windows on a cold and windy night in your bedroom, you are not allowing for passive ventilation to appropriately exchange the air in your home.

That said, there are many times of the year where passive strategies will work well, and designing them into the building as a resilience measure in the event that there is power loss or mechanical failure is an appropriate measure. "Night flushing" is a cooling strategy commonly employed in regions where nighttime temperature drops below indoor temperatures; one simply opens the windows at night to encourage a stack effect in the building, "flushing" out the hot air from the day and cooling the home down.

Domestic Hot Water (DHW) and Plumbing

For a high-performance home, we must consider the DHW load not only as carefully as the heating load, but possibly as the *largest* load in the building, surpassing heating by a factor of two or greater if we have done a good job in reducing our heating loads. Accordingly, additional design time is warranted to appropriately size the DHW system and take measures to reduce water — and therefore energy — consumption. There are many considerations in good DHW design, many of which parallel the considerations for heating systems explored earlier (i.e., fuel type, integration to other HVAC systems); however, some considerations are unique to water systems (i.e., hardness, pH, turbidity). As potentially the single largest energy load in the building, care should be taken in designing the DHW system to align with the goals and intentions set forth for the rest of the design program.

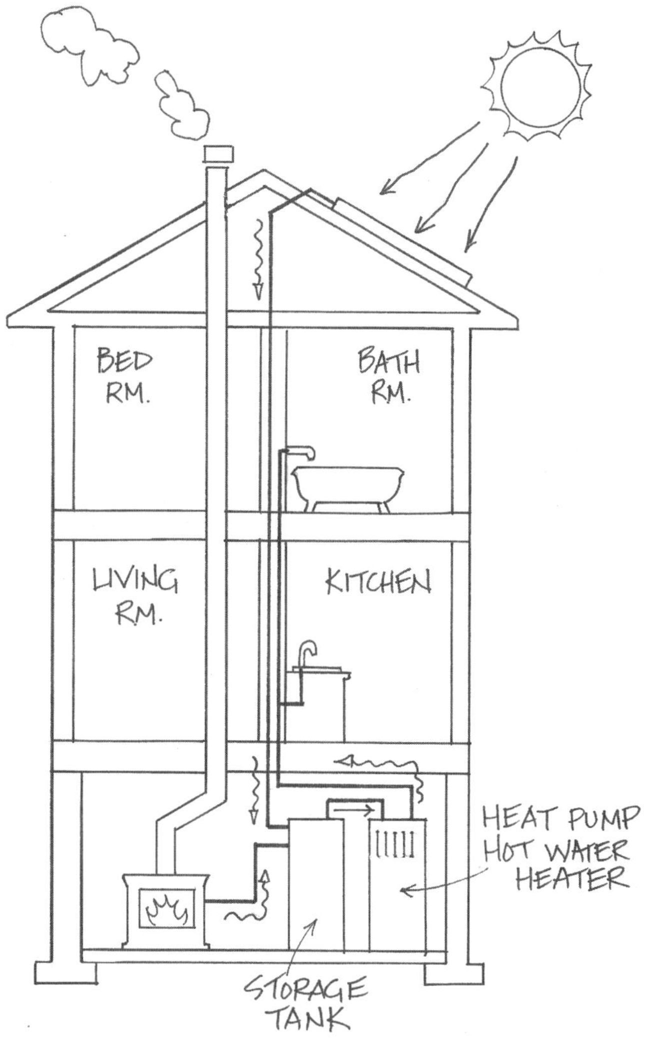

Fig. 7.9: *In most high-performance buildings, the DHW is the largest single load in the building, and attention should be paid in design to ensure good efficiency is realized without sacrificing functionality.*

Considerations around plumbing are very practical and generally relate to moisture concerns, yet they are often overlooked:

- Keep all live plumbing out of insulated exterior assemblies and to the interior of the enclosure when possible to avoid freezing, allow for maintenance, and protect sensitive enclosure materials from leaks and the resulting damage.
- Hot water tanks and clothes washers are particularly leak-prone, and should be located in rooms with drains and moisture-resistant flooring. Provide accessible shut-off valves to facilitate maintenance and replacement.
- Cold pipes are susceptible to condensation; insulate both hot and cold pipes to avoid condensation and improve efficiency.
- Enclosure spaces behind bathtubs and showers must be fully sealed with moisture-durable materials — do not leave exposed cavities or use paper-faced wallboard or other mold-prone materials behind these built-in fixtures.

Occupants: Controls and Management

The last factor in the "whole system" of the building, both chronologically and functionally, is the occupant; while the fundamentals of the design, construction, and installation of the enclosure and MEP systems must be in place for a high-performance home to exist, it is the long-term operation of the home that defines the success of the project. There is an adage that states there are no Net Zero or PassiveHouse buildings, only Net Zero and PassiveHouse occupants in buildings that allow them to achieve these goals. There is much that has been written about the role of occupants in supporting — or thwarting — the design intentions of a building. For now, we will simply identify some of the key control and management components that should be considered regarding the homeowner's interface with the technological systems of their home.

Simple programmable thermostats will ensure the greatest likelihood of hitting the "sweet spot" of maximum comfort and efficiency; the advent of new "learning" thermostats that automatically take in a wide variety of inputs from the exterior and interior climates and the operation of the mechanical equipment can further simplify owner interface, allowing a "set it and forget it" approach that doesn't sacrifice performance or comfort. For owners who desire greater monitoring and control, many modern Internet of Things (IoT) HVAC controls offer remote operation of the building via a smartphone app.

Ventilation controls are similar in many ways to heating and cooling controls — the priorities are simple, programmable interfaces that allow for good baseline operation of the system; in most cases, whole-house ventilation controls are even less noticeable (and therefore demand less occupant control) than heating and cooling controls. Spot ventilation is the outlier here, however, and the ability to ensure good ventilation control is imperative. Humidistats and motion controls are now standard for many bath fans; HRVs and whole-house exhaust bath fans often come with timed "boost" controls to allow for targeted spot ventilation without compromising the automated dilution ventilation functionality. Advanced systems often feature carbon monoxide (CO) and volatile organic compound (VOC) sensors to drive automated ventilation control.

Electric load controls continue to improve. There is a wide range of strategies being developed, including "smart" power strips and switched outlets for phantom load control (especially useful for entertainment and other "always on" remote-activated appliances and electronics), and adjustable, timed, and motion control for lighting.

Monitoring and maintenance of energy loads in the building is one of the least-employed and most-valuable features in the modern home. The simplest version of this is an Energy Use Intensity (EUI) calculation, in which the total energy load of a building (easily calculated from the utility and fuel bills) is divided by the building square footage to get a quick measure of how well the building is performing. Energy and HVAC "dashboard" or other monitoring programs can support long-term maintenance and operation of the many components of our buildings' systems. Of course, fundamental to the long-term performance of the building systems is the initial selection of low-maintenance and long-lasting appliances and placing them in accessible locations for easy servicing.

Finally, providing a quality owner's manual with a new high-performance home should be standard practice for all new construction. This need not be overly complicated, but taking the time to compile critical information regarding the building enclosure and systems (including their ongoing repair and maintenance requirements) will go a long way to ensuring success, even after the building gets sold to new owners. Basic information might include "as-built" plans, a portfolio of all the manufacturer's instruction booklets that came with appliances and fixtures, instructions on how to repair or seal new work to the air barrier, vapor considerations regarding new paints and finishes, an explanation and schematic of the basic MEP systems and listing of their parts, maintenance schedules, and part vendors. It's also a good idea to include a list of the critical subcontractors names and emergency contact information. A bonus feature would be a report outlining the results of the commissioning of the HVAC systems. This does two important things: it verifies that all the systems were checked to ensure they were working as designed, and it gives future contractors a baseline to refer to when something goes wrong. While this may cost you some extra time, the value is beyond measure.

Remember: the systems of a building will only function as well as their owners' ability to operate them, and it is our responsibility to set the owners up for success.

Endnotes

Chapter 1

1. Straube, John. "The Perfect HVAC," BSI-022, Building Science Corporation, 6/3/11, buildingscience.com
2. "Scientific Consensus: Earth's Climate Is Warming," NASA, climate.nasa.gov
3. "Roadmap to Zero Emissions," Architecture 2030, June 4, 2014, unfccc.int

Chapter 2

1. Straube, John. "How Heat Moves Through Homes — Building Science Podcast," GreenBuildingAdvisor.com, 4/12/10, greenbuildingadvisor.com
2. Straube, John. "Thermal Metrics for High Performance Enclosure Walls: The Limitations of R-Value," Research Report 0901, Building Science Corporation, 2007, buildingscience.com
3. Kosny, Jan, Ph.D. "A New Whole Wall R-Value Calculator," Oak Ridge National Laboratory Buildings Technology Center, 08/04, web.ornl.gov
4. Straube, John. "Thermal Metrics."
5. Ibid.
6. Ibid.
7. Straube, John. "Air Flow Control in Buildings," BSD-014, Building Science Corporation, 10/15/07, buildingscience.com
8. Wilson, Alex. "Storing Heat in Walls with Phase-Change Materials," GreenBuildingAdvisor.com, 11/24/09, greenbuildingadvisor.com.
9. Holladay, Martin. "All About Thermal Mass," GreenBuildingAdvisor.com, 5/3/13, greenbuildingadvisor.com
10. Ibid.
11. Ibid.

Chapter 3

1. Lstiburek, Joseph. "Insulations, Sheathings and Vapor Retarders," Research Report 0412, Building Science Corporation, 11/04, buildingscience.com
2. Ibid.
3. Ibid.
4. Ibid.
5. Lstiburek, Joseph. "Solar Driven Moisture in Brick Veneer," RR-0104, Building Science Corporation, 9/15/01, buildingscience.com
6. Lstiburek, Joseph. "Chubby Checker and the 'Fat Man' Do Permeance," BSI-087, Building Science Corporation, 6/15/15, buildingscience.com

Chapter 4

1. Kung'U, Jackson. "Factors that Affect the Growth of Moulds," Mold and Bacteria Consulting Services (MBS), 2016, moldbacteriaconsulting.com
2. Anagnost, Susan. "Wood Decay, Fungi, Stain and Mold," New England Kiln Drying Association Meeting, April 7, 2011, Onconta, New York, esf.edu
3. John Straube in Bruce King, *The Design of Straw Bale Buildings: The State of the Art.* p. 144. San Rafael, CA: Green Building Press, 2006.
4. Lstiburek, Joe. "Water Management," RR-0103, Building Science Corporation, September 15, 2001, buildingscience.com
5. Air Barrier Association of America. "About Air Barriers: Materials, Components, Assemblies, & Systems," n.d., accessed June 26, 2016, airbarrier.org
6. Straube, John. "Thermal Metrics for High Performance Enclosure Walls: The Limitations of R-Value," Research Report 0901, Building Science Corporation, 2007, buildingscience.com

7. Lstiburek, Joseph. "Understanding Vapor Barriers," BSD-106, Building Science Corporation, April 15, 2011, buildingscience.com
8. "Vapor Retarder Classification," Building America Top Innovations PNNL-SA-90571, US DOE, January 2013, apps1.eere.energy.gov
9. "Seasonal Performance of Passive Radon-Resistant Features in New Single-Family Homes," NAHB Research Center, Inc. for US EPA, May 1996, nchh.org

Chapter 5

1. *Insulation Recommendations: A Quick Guide to Cost, Health, and Environmental Considerations.* BuildingGreen, 2013–2016, buildinggreen.com,
2. Holladay, Martin. "In Cold Climates, R-5 Foam Beats R-6," 12/13/15, greenbuildingadvisor.com
3. White, David. "Insulation GWP Tool v 1.2, Right Environments," 2011, rightenvironments.com
4. "Energy Ratings: What the Ratings Mean," National Fenestration Research Council, 2012, nfrc.org
5. Holladay, Martin. "All About Glazing Options," updated May 5, 2016, greenbuildingadvisor.com
6. BizEE DegreeDays.Net, BizEE Software Ltd., Uplands, Swansea, UK, accessed 5/13/16, degreedays.net
7. Bailes, Allison. "Do You Know Your Building Science Climate Zone?," Energy Vanguard, 4/22/13, energyvanguard.com
8. Wilson, Alex. "How Much Insulation is Needed?," July 15, 2009, greenbuildingadvisor.com
9. Holladay, Martin. "Can Houses be 'Too Insulated' or 'Too Tight'?" accessed May 12, 2016, greenbuildingadvisor.com
10. Straube, John. Private correspondence, 2012. University of Waterloo, Ontario.
11. Building Energy Software Tools Directory, International Building Performance Simulation Association — USA, accessed 5/13/16, buildingenergysoftwaretools.com

Chapter 6

1. Bailes, Allison. "What Is the Best Way to Deal with Crawl Space Air?," Energy Vanguard, September 16, 2013, energyvanguard.com
2 "Revised Builder's Guide to Frost Protected Shallow Foundations," NAHB Research Center, September 2004, homeinnovation.com
3. Lstiburek, Joseph. "The Perfect Wall," BSI-001, Building Science Corporation, July 15, 2010, buildingscience.com/documents/insights/bsi-001-the-perfect-wall
4. Holladay, Martin. "Calculating the Minimum Thickness of Rigid Foam Sheathing," GreenBuildingAdvisor.com, updated 2/26/16, greenbuildingadvisor.com
5. Holladay, Martin. "How to Install Rigid Foam Sheathing," GreenBuildingAdvisor.com, updated 7/20/15, greenbuildingadvisor.com
6. "Flow-Through Assemblies," BSI-091, Building Science Corporation, January 13, 2016, buildingscience.com
7. Ibid.
8. "Water Management Details for Residential Buildings," BSP-020, Building Science Corporation, January 2007, buildingscience.com
9. Holladay, Martin. "All About Rainscreens," GreenBuildingAdvisor.com, updated 6/17/13, greenbuildingadvisor.com
10. Lstiburek, Joseph. "Roof Design," RR-0404, Building Science Corporation, 9/15/04, buildingscience.com
11. Lstiburek, Joseph. "Unvented Roof Systems," RR-0108, Building Science Corporation, 1/15/01, buildingscience.com

12. Lstiburek, Joseph. "Venting Vapor," BSI-088, Building Science Corporation, 7/15/15, buildingscience.com

Chapter 7

1. Lstiburek, Joseph. "Read This Before You Design, Build, or Renovate," Building Science Corporation, revised May 2005, buildingscience.com
2. "The Difference Between Source and Site Energy," US DOE EnergyStar, energystar.gov
3. Deru, M. and Torcellini, P., "Source Energy and Emission Factors for Energy Use in Buildings" Technical Report NREL/TP-550-38617, Revised June 2007.
4. "Table: Whole Building Ventilation Requirements," Standard 62.2 2013: Ventilation and Acceptable Indoor Air Quality in Low-Rise Residential Buildings, ASHRAE, 2013, ashrae.org
5. Lstiburek, Joseph. "Read This Before You Design."
6. Sterling, E.M., A. Arundel, T.D. Sterling. "Criteria for human exposure to humidity in occupied buildings," ASHRAE Transactions, Vol. 91, Part 1B, 1985, pp. 611–622.

Resources

Governments

U.S. Environmental Protection Agency
ENERGY STAR Buildings Program & Indoor Environments Division
1200 Pennsylvania Avenue, NW
Washington, DC 20460
888.STAR.YES
epa.gov/iaq and energystar.gov

U.S. Department of Energy,
Building America Program
1000 Independence Ave., SW
Washington, DC 20585
800.dial.DOE
eren.doe.gov

Oak Ridge National Laboratory
U.S. Department of Energy
P.O. Box 2008
Oak Ridge, TN 37831
865.576.7658
ornl.gov

Canada Mortgage and Housing Corporation
700 Montreal Road
Ottawa, Ontario K1A 0P7
613.748.2000
cmhc-schl.gc.ca

National Resources Canada
343.292.6096
nrcan.gc.ca/energy/efficiency/housing

Non-profits

Home Performance Coalition
One Thorn Run Center
1187 Thorn Run Extension, Suite 340
Moon Township, PA 15108
800.344.4866
homeperformance.org

Energy & Environmental Building Association
10740 Lyndale Avenue South, Suite 10W
Bloomington, MN 55420-5615
952.881.1098
eeba.org

Rocky Mountain Institute
1739 Snowmass Creek Road
Snowmass, CO 81654-9199
970.927.3851
rmi.org

National Institute for Building Science
1090 Vermont Avenue, NW, Suite 700
Washington, DC 20005-4950
202.289.7800
nibs.org

Architecture 2030
607 Cerrillos Road
Santa Fe, NM 87505
505.988.5309
architecture2030.org

Canada Green Building Council (CaGBH)
47 Clarence Street, Suite 202
Ottawa, Ontario K1N 9K1
866.941.1184
cagbc.org/

Publications — Online

Green Building Advisor
Taunton Press
greenbuildingadvisor.com

Publications — Forum

BuildingGreen/Environmental Building News
buildinggreen.com

Whole Building Design Guide
National Institute of Building Science
wbdg.org

Publications — Periodicals

Journal of Light Construction
Hanley Wood Media
One Thomas Circle, NW
Suite 600
Washington, DC 20005
202.452.0800
jlconline.com

Fine Homebuilding
Taunton Press
63 South Main St., PO Box 5506
Newtown, CT 06470-5506
203.426.8171
finehomebuilding.com

Home Energy magazine
1250 Addison Street
Suite 211B
Berkeley, CA 94702
510.524.5405
homeenergy.org

Publications — Print

Lstiburek, Joseph: EEBA Builder's Guide series
Building Science Press
Westford, MA, 2009–2012.

Rose, William: *Water in Buildings: An Architect's Guide to Moisture and Mold.*
Wiley and Sons, 2005.

Other Resources

Building Science Corporation
70 Main Street, Westford, MA 01886
978.589.5100
buildingscience.com
(publications, training, technical assistance, design)

NAFS — North American Fenestration Standard
AAMA/WDMA/CSA 101/I.S.2/A440-08
aamanet.org/upload/file/CMB-5-08.pdf

Index

Page numbers in *italics* indicate tables.

About the Author

Jacob Deva Racusin is a sustainable and natural building designer, builder and educator. He is co-author of *The Natural Building Companion*, contributor to *The Art of Natural Building* and Systems Director and Co-Owner of New Frameworks Natural Design/Build, focusing on mechanical, water, energy, and enclosure system design and quality control. He is also a Building Performance Institute-certified Envelope Professional and Building Analyst. Jacob is the program director of the Building Science and Net Zero Design Certificate Program at the Yestermorrow Design/Build School, and has taught natural building and building science at various universities and building schools. He and his family live in a 2000 sq ft high-performance, natural home in the mountains of northern Vermont, where they run a small-scale Permaculture-inspired homestead.

A Note About the Publisher

NEW SOCIETY PUBLISHERS (**www.newsociety.com**), is an activist, solutions-oriented publisher focused on publishing books for a world of change. Our books offer tips, tools, and insights from leading experts in sustainable building, homesteading, climate change, environment, conscientious commerce, renewable energy, and more — positive solutions for troubled times.

Sustainable Practices for Strong, Resilient Communities

We print all of our books and catalogues on **100% post-consumer recycled paper**, processed chlorine-free, and printed with vegetable-based, low-VOC inks. These practices are measured through an Environmental Benefits statement (see below). We are committed to printing all of our books and catalogues in North America, not overseas. We also work to reduce our carbon footprint, and purchase carbon offsets based on an annual audit to ensure carbon neutrality.

Employee Trust and a Certified B Corp

In addition to an innovative employee shareholder agreement, we have also achieved B Corporation certification. We care deeply about *what* we publish — our overall list continues to be widely admired and respected for its timeliness and quality — but also about *how* we do business.

For further information, or to browse our full list of books and purchase securely, visit our website at: **www.newsociety.com**

New Society Publishers

ENVIRONMENTAL BENEFITS STATEMENT

For every 5,000 books printed, New Society saves the following resources:[1]

34	Trees
3,090	Pounds of Solid Waste
3,400	Gallons of Water
4,435	Kilowatt Hours of Electricity
5,618	Pounds of Greenhouse Gases
24	Pounds of HAPs, VOCs, and AOX Combined
9	Cubic Yards of Landfill Space

[1]Environmental benefits are calculated based on research done by the Environmental Defense Fund and other members of the Paper Task Force who study the environmental impacts of the paper industry.